Raspberry Pi with Custard

Interfacing to the Raspberry Pi GPIO

Seggy T Segaran

First published in 2014 by Ohm Books Publishing

ISBN-13: 978-1482098143

© 2014 Seggy T Segaran
All rights reserved.
The moral right of the author has been asserted.

This book is sold subject to the condition that it shall not, by way of trade or otherwise, be lent, re-sold, hired out, or otherwise circulated without the publishers prior consent in any form of binding or cover other than that in which it is published and without a similar condition including this condition being imposed on the subsequent purchaser.

CONTENTS

Introduction		4
Chapter 1	Getting Started	5
Chapter 2	Custard Pi 1 - Breakout Board with protection	17
Chapter 3	Custard Pi 2 - General Purpose input/output board	28
Chapter 4	Custard Pi 3 - 8 analogue input board	40
Chapter 5	Custard Pi 5 - Breakout board for digital 8 I/O	49
Chapter 6	Custard Pi 6 - 8 Relay Card	54
Chapter 7	Custard Pi 7 - Industrial Control Interface	68
Chapter 8	Custard Pi 8 - SMD and Through Hole Breadboard	91
Chapter 9	Custard Pi Casing and Base	93
Chapter 10	Custard Pi Combos	95
Appendix A	Starting up Desktop on Power-up	100
Appendix B	Auto-starting a Python Program on Power-up	101
Appendix C	Comparison of Raspberry Pi Models	102

Introduction

If you are interested in tinkering with electronics and have not heard of the Raspberry Pi, then you have probably been on an extended holiday with no access to the internet. The product of the Raspberry Pi Foundation, this is a credit card sized £25 computer that has caught people's imaginations. It has sold more than a million and is being used in a wide variety of applications.

Like the Sinclair ZX80 and the Acorn Atom of the 1980s, the Pi has an open architecture that makes it extremely easy for users to program. More importantly there are no licensing fees for the software that is used.

This book is specifically about interfacing the Pi to the external world, using the General Purpose Input Output (GPIO) port that is built onto the board using the range of Custard Pi plug-in cards. The reader can learn how to switch things on and off, read the position of switches and sensors and take measurements of temperature.

This book is not a detailed guide on how to program the Pi. There is an introduction to the Python programming language and sample code to drive particular bits of external hardware. It assumes that the average reader has some programming experience.

If you are a programmer who wants to use the GPIO port to interface to the real world then this book is for you. It enables you to have some real fun with the Raspberry Pi.

If you are interested in using electronics in a fun practical way then this book will help you. Learning maths in a classroom is no way to learn about electronics. The best way is to build something that works by tinkering. The maths gives you the background but lacks passion and the gratification of building something that works.

Have fun with this book.

Seggy T Segaran

CHAPTER 1   GETTING STARTED

1.1   Raspberry Pi Basics

The story of the Raspberry Pi is story about a great British engineering success. In the words of Eben Upton, the founder of the Raspberry Pi foundation:

"Originally conceived as a fun easy way for kids (and curious adults) to learn computer programming, the Raspberry Pi quickly evolved into a remarkable robust credit-card-sized computer that can be used for everything from playing HD videos and hacking around with hardware to learning to program!"

The original plan was to build a few thousand units over the lifetime of the project. Due to the price point of £25 and the huge interest generated in this project by the media, sales have exceeded a million as this book is being written.

There has been much written about the Raspberry Pi over the last year, both on the Internet and in books. There seems little point in me repeating much of it here, so I shall keep this chapter brief.

Credit card sized computer motherboard

This is a good description of the Raspberry Pi and is a useful way to understand what it is, as most people are familiar with desk top PCs and laptops. We highlight the major differences below.

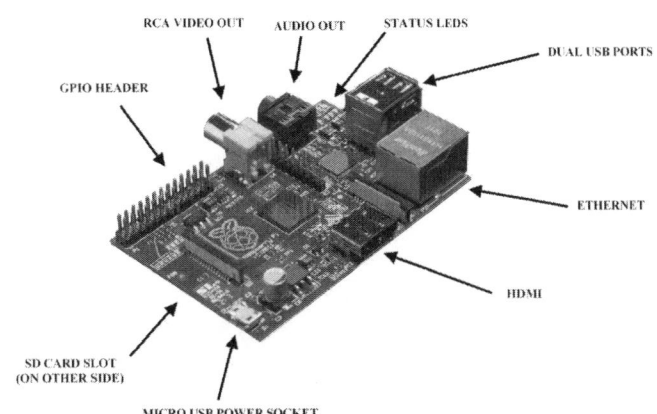

Raspberry Pi Model B with components shown

Size: A PC motherboard can be 20cm x 30cm while the Raspberry Pi is about 6cm x 9cm. That is 10 times smaller in area.

Power: The Raspberry Pi uses a 5W power supply. A PC uses at least 50W.

Memory: A PC has a hard disk for storage. The Raspberry Pi uses an SD card.

Operating system: Most PCs use Windows and a license fee has to be paid to Microsoft. The Raspberry Pi uses Linux, which is open source.

Hacking: PCs are getting slicker all the time and normal users are discouraged from "looking under the bonnet". It is virtually impossible to get inside a laptop enclosure. The Raspberry Pi encourages experimentation and is supplied without an enclosure.

Connecting to the outside world: With a PC, there are set ports used for this, such as USB. With the Raspberry Pi, there is a GPIO port to connect to the outside world.

Programming: Although there is wide variety of programming languages available for a PC, it is not easy for a newcomer to get started with this. The operating system is geared towards end user applications like word processing, spreadsheet and databases. As soon as the Raspberry Pi is powered up, one can start using Python. The Pi in Raspberry Pi refers to Python which is an easy to learn programming language.

Before you can start

To keep costs down, the Pi is supplied as an uncased assembly. To get it working, you will need the following accessories.

USB keyboard and mouse

Power supply (5V, 1 Amp) with micro USB connector

HDMI cable to TV

SD card with Linux operating system

Enclosure

All of the above are widely available from various internet retailers. The last of these is optional, but advisable. Electronic circuits are liable to damage from static electricity built up in the body and then discharged through to components. There is also the danger of accidental shorting on conductive components such as paper clips left on a table. So my advice is to put your Raspberry Pi in a case as soon as you can.

The power supply needs to be capable of sourcing 1 Amp to power the Pi board as well as the USB keyboard and mouse. When the Raspberry Pi boots up the green led (next to the blue audio connector) starts flashing. If this does not happen then there is likely a problem with the Power supply. Some of the USB power supplies on the internet are not capable of supplying 1 Amp, even though they are marked as such. It pays to source these components from official Raspberry Pi stockists like Farnell or RS Components.

The Raspberry Pi does NOT come with user instructions either. I would recommend that users wanting get most out of the Pi should invest in the book by Eben Upton "Raspberry Pi User Guide". Not only does this book present the basics of the Pi, it has a useful introduction into the history of the Raspberry Pi Foundation and its objectives.

Raspberry Pi Model B+

This model was launched in 2014. The main differences that are of interest to us are as follows.

	Model B	**Model B+**
GPIO connections	26	40
USB ports	2	4
Storage	SD card	Compact SD card
Stereo/video connectors	Two connectors	Single connector
Mounting holes	2	4

1.2 Getting to the Raspberry Pi on the Screen

Plug the keyboard and mouse into the dual USB sockets of the Raspberry Pi. (The Model B comes with dual USB sockets. The Model A has only one USB socket and you will need a USB hub to allow you to connect both keyboard and mouse in.

Connect to the television using the HDMI lead and insert the SD card with the Linux operating system. Be careful when inserting the SD card.

SD card in slot on underside of the board (Model B)

The SD card slot on the Raspberry Pi is quite delicate and I have managed to break 2 of these. Looking on the internet, it is clear that this has happened to a lot of other users as well. Appendix A shows a way of fixing broken SD card slots and also a way of protecting them from damage.

Raspberry Pi all connected up

Connect the power supply to the mains and plug into the micro USB socket on the Pi. The LEDs next to the blue audio socket should start to flash you should see text scrolling on the TV. After a little while, you should get a line saying "raspberry pi login:".

Enter "pi" Then you should then get a line saying "Password". Enter "raspberry". After a few lines the following command line prompt should appear: pi@raspberrypi ~$ followed by a flashing cursor.

Enter "startx" to launch the graphical user interface (GUI) shown on the next slide.

At this point, if you have the Raspberry Pi connected to your home internet hub, you can access the web by double clicking on the Midori icon. This is the internet browser supplied on the SD card.

1.3 GPIO port

It's great to have a computer to do basic office related stuff like word processing, spreadsheets and also to access the Internet for searching and e-mails. However it's enormous fun to be able to control or monitor external devices. As computers have evolved from the early Acorn Atoms and BBCs to the IBM PCs and Macs, they have shed the ability to easily control external hardware. On a PC the only way one can do this is via the proprietary USB ports. It is no longer possible to hack around with simple bits of hardware. One has to use USB drivers, and external USB to relay cards to be able to do anything. The user has been "locked out" of the pleasures of simply messing around with electronics hardware.

Well, the Raspberry Pi has blown a hole through this and provided us with an array of pins bundled together as the GPIO port.

The GPIO port on the Model B

This port has 26 pins and the layout of this is shown below.

Function	Pin Numbers		Function
3.3V	1	2	5V
I2C SDA	3	4	5V
I2C SCL	5	6	Gnd
GPIO 4	7	8	UART TXD
Gnd	9	10	UART RXD
GPIO 17	11	12	GPIO 18
GPIO 27	13	14	Gnd
GPIO 22	15	16	GPIO 23
3.3V	17	18	GPIO 24
SPI MOSI	19	20	Gnd
SPI MISO	21	22	GPIO 25
SPI CLK	23	24	SPI CE0
Gnd	25	26	SPI CE1

Layout of GPIO port pins (rev 2 board)

Power pins

The GPIO port has the 3.3V on pin 1 and 5V on pin 2. This can be used to power external electronics as long as not too much current is drawn. We would advise that a maximum of 50mA can be drawn from the 3.3V pin and a maximum of 100mA from the 5V rail when the Raspberry Pi is being powered from a 5V 1A supply. The 0V (or Gnd.) connection is on pin 6.

On revision 1 boards there should be no connections made to pins 4, 9, 14, 17, 20 and 25.

General Purpose Input Output pins

The pins marked green are general purpose digital input output pins. These are pins 11, 12, 13, 15, 16, 18, 22 and 7. They can be set high (to 3.3V) or low (0V) by program control from the Raspberry Pi. The Pi can also read whether these pins are high or low, say due to a switch being pressed.

I2C serial bus

This bus is used to interconnect various integrated circuits such as A to D and D to A convertors, Port expanders and tone generators. The bus has a bi-directional data line (SDA on pin 3) and a clock line (SCL on pin 5). The I2C bus can interface to many devices connected to the same 2 lines as the devices are addressable individually.

SPI serial bus

This serial bus can also be used to interface to a number of external integrated circuits. However it is different from the I2C bus in that it has separate data out (MOSI on pin 19) and data in (MISO on pin 21) lines and the devices are not addressable. There are separate chip enable lines for each integrated circuit. The Raspberry Pi SPI bus is provided with 2 chip enable outputs (CE0 on pin 24 and CE1 on pin 26).

UART serial port

This is a serial port and can be used to communicate with external devices equipped with an RS232 serial interface. The transmit output (TXD) is on pin 8 and the receive input (RXD) is on pin 10.

1.4 Using Python

Launch IDLE3 from the Raspberry GUI. This will allow us to experiment with Python commands.

At the >>> prompt type

```
print ('hello world')
```

You will see "hello world" on the screen.

At the >>> prompt type

```
20 + 5
```

You will see "25" on the screen.

This is a way of executing Python commands immediately. A programme is simply a collection of commands executed consecutively.

Writing a Python program

Under the file tab in IDLE3, click on new window. You are now ready to write a program. For now type the following exactly.

```
For x in range (1,10);
    y=x * x
    print "x=", x, "square of x=", y
```

Under the file tab, click on save and call the file "squarex".

Launch LXTerminal from the Raspberry Pi GUI. This will bring you back to the Linux command line prompt: pi@raspberrypi - $

Type "`sudo python squarex.py`" to run your program.

You will see
X = 1 square of x = 1
X = 2 square of x = 4
...
X = 9 square of x = 81

That's it. You've written your first Python program.

Some notes on Python

You can run programs using the "run module" option under the run tab. However when I tried this I got programming syntax errors. When I ran the same program using the command line prompt, it worked fine.

Sudo – stands for "super user do". With Linux, this gives you the right privileges to run a Python program.

Comments - If you want to add comments to your program, then use # at the start. For example, you could have started the program on the previous page with

`# program to work out squares of numbers from 1 to 9.`

Using libraries – Many standard functions are available as libraries in Python. These can be used by using the "import" command and will save you a lot of time in programming.

For example "import time" will bring in a library which can be used for time delays. "import random" will bring in a random number generator.

Turning an led on and off using Python code

We are going to use the GPIO port on the Raspberry Pi for this. To make the connection easier, the Custard Pi 1 breakout board is used. This plugs straight into the GPIO connector, provides easy screw terminal connection and protects the Raspberry Pi from accidental damage.

14

Raspberry Pi with Custard

We are using pin 11 of the GPIO port. This is available on connector J2 of the Custard Pi 1 and is labelled as pin 11. The 0V (or Gnd) connection is the centre pin of the 3 pin power connector J3.

Connecting the led

When pin 11 is True (taken high) the voltage on it will be almost 3.3V. This needs to go to the positive side of the led. This is the longer leg of the led. The other side of the led goes to 0V (Gnd).

Note: The Custard Pi 1 board has a 220 ohm resistor on each pin so you can connect this to the LED. If you are not using the Custard Pi 1 then you have to use an external resistor of around 330 ohm in series with the LED.

Python program to flash led

Type in the program steps below. The text behind the # explains what the code does, but you do not have to enter this.

```
Import RPi.GPIO as GPIO      #import GPIO library
Import time                  #import time library
GPIO.setmode(GPIO.BOARD)     #use board pin numbers
GPIO.setup(11, GPIO.OUT)     #setup pin 11 as output
For x in range (0,10):       #repeat for x=0 to 9
   GPIO.output(11, True)     #set pin 11 high
   time.sleep(0.2)           #wait 0.2 seconds
   GPIO.output(11, False)    #set pin 11 low
   time.sleep(0.2)           #wait 0.2 seconds
GPIO.cleanup()               #tidy up GPIO port
Import sys                   #exit program
Sys.exit()
```

Save the file as "ledonoff.py".

Trying out the program

15

Open LXTerminal and type "`sudo python ledonoff.py`" to run the program. The led should flash ten time at a fairly fast rate.

Try changing the time.sleep line from 0.2 seconds to 0.5 seconds. When you run the program, it should flash ten times, but fairly slowly this time.

Tip: To rerun the program, press the upwards arrow ↑ to re-enter the last command on the screen and then press return to run it.

Now change the x in range command from 0,10 to 0,5 to flash the led just 5 times.

Well done. You have just managed to write some Python code to flash an LED.

Raspberry Pi with Custard

CHAPTER 2 CUSTARD PI 1
Breakout Board with protection for the Raspberry Pi GPIO

2.1 Introduction

The Raspberry Pi GPIO allows the control of external electronics. There are two rows of 13 pins which are brought out to a 26 way header on the edge of the board. The Custard Pi 1 board simply plugs into the Raspberry Pi GPIO connector and allows users to quickly connect to all the pins. At the same time it protects the Raspberry Pi from possible damage from the wrong voltage being accidentally connected to the GPIO.

Custard Pi 1 plugged into the Raspberry Pi GPIO

Custard Pi 1 plugged into Raspberry Pi Model B+

2.2 Circuit Description

When the Custard Pi is plugged into the GPIO, two LEDs come ON, showing that the 5V and 3.3V rail are working correctly.

17

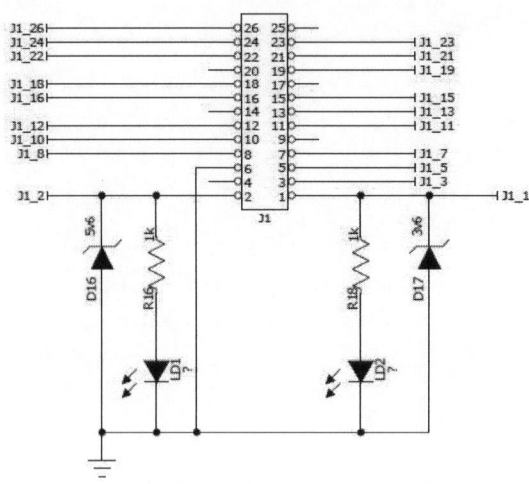

GPIO connections showing power rails

The 3.3V is supplied on pin 1 of the GPIO and the 5V is supplied on pin 2. The 2 LEDs are connected to these pins with a 1k current limiting resistor. Note that there are no connections to pins 4, 9, 14, 17, 20 & 25 of the GPIO.

The other pin connections are outlined below.

Pin 1: 3.3V supply
Pin 2: 5V supply
Pin 6: 0V
Pins 11, 12, 13, 15, 16, 18, 22 and 7: Digital input output pins
Pins 3 and 5: I2C bus
Pins 19, 21, 23, 24 and 26: SPI bus
Pins 8 and 10: UART

Power pins (J3)

These are brought out to connector J3 on the Custard Pi 1 and have a fuse fitted to each line. This is to prevent the user from drawing too much current from the Raspberry Pi. The fuses are resettable and are both rated at 0.1 Amp (100 m Amp).

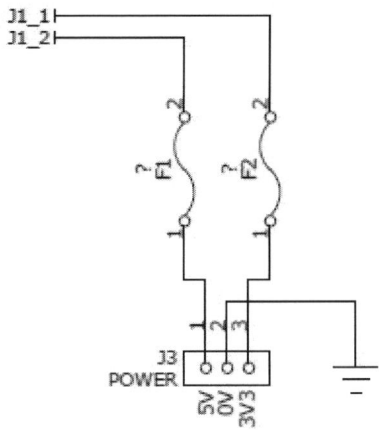

5V and 3.3V pins with fuses

General Purpose Input Output (I/O) pins (J2)

The pins marked green are general purpose digital input output pins. These are pins 11, 12, 13, 15, 16, 18, 22 and 7. They can be set high (to 3.3V) or low (0V) by program control from the Raspberry Pi. The Pi can also read whether these pins are high or low, say to a switch being pressed.

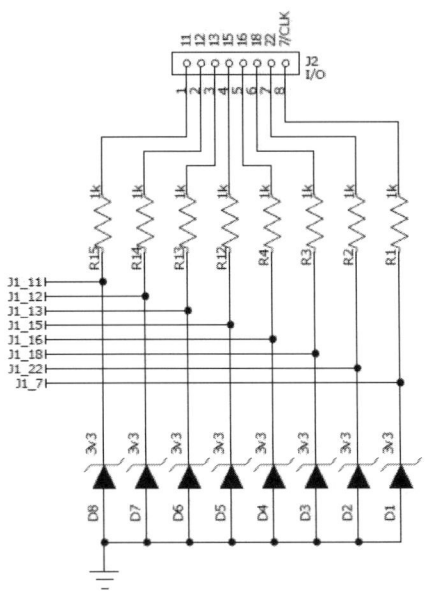

General purpose I/O pins

Each of these pins is protected by a 3.3V zener and a series current limiting resistor. The 3.3V zener prevents any voltages in excess of 3.3V from being applied to the pins of the Integrated Circuit on the Raspberry Pi and damaging it. It also protects from negative voltages being applied to the pins. Due to the diode action of the zener voltages on the pins are limited to -0.7V.

The 1k ohm (1000 ohm) current limiting resistors are there to prevent too much current flowing in the zener if a wrong voltage was to be connected. However on later Custard Pi 1 boards this has been reduced to 220 ohms. The 1000 ohm resistor has colour bands brown, black red and the 220 ohm resistor has colour bands red, red and brown.

I2C serial bus (J7)

This bus is used to interconnect various integrated circuits such as A to D and D to A convertors, Port expanders and tone generators. The bus has a bi-directional data line and a clock line and can interface to many devices connected to the same 2 lines as the devices are addressable individually.

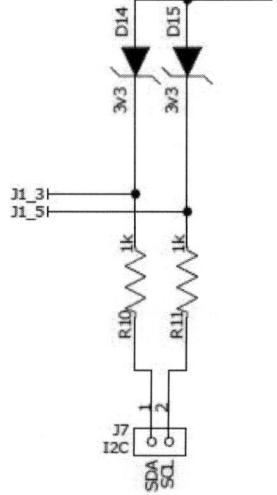

I2C bus pins

Internally, the Raspberry Pi uses two pull up resistors (1.8 k ohms) on these 2 pins. If you are planning to use the I2C bus then make sure that the current limiting resistors are 220 ohm and not 1000 ohms. If you have a Custard Pi 1 board with a 1k ohm resistor fitted, then either replace this with a 220 ohm resistor or solder a 220 ohm resistor on top of the 1k ohm resistor.

SPI serial bus (J5)

This serial bus can also be used to interface to a number of external integrated circuits. However it is different from the I2C bus in that it has separate data out and data in lines and the devices are not addressable. There are separate chip enable lines for each integrated circuit. The Raspberry Pi SPI bus is provided with 2 chip enable outputs.

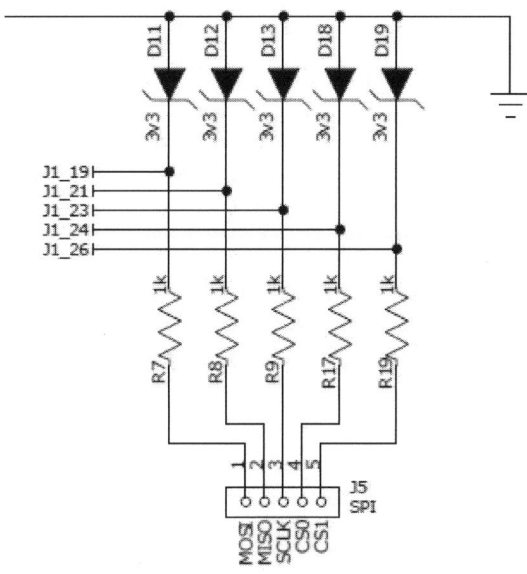

SPI bus pins

UART serial port (J6)

This is a serial port and can be used to communicate with external devices equipped with an RS232 serial interface.

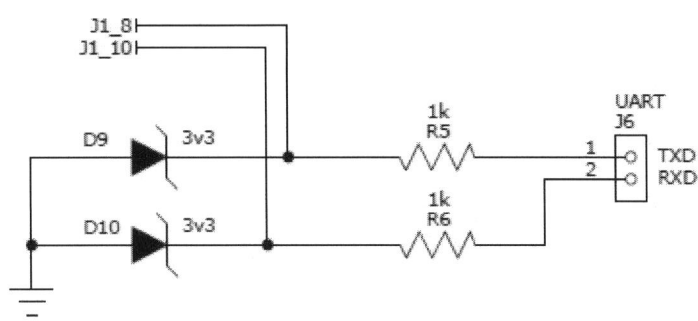

UART pins

Raspberry Pi with Custard

2.4 The Custard Pi 1 Assembly

The positions of the connectors are shown below. These are mini screw terminals into which wires can be quickly connected.

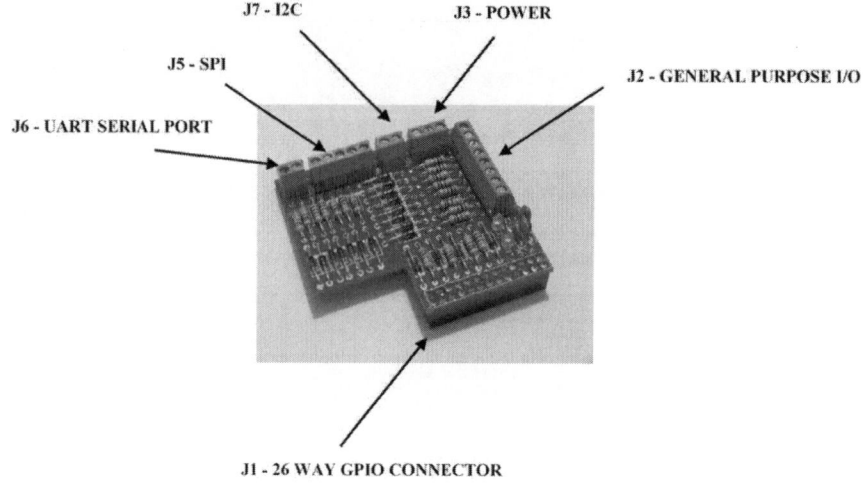

Positions of the connectors

This is a compact assembly that simply plugs into the Raspberry Pi GPIO. This can be done even with the Raspberry Pi is powered. Just make sure that the 2 power LEDs are on as soon as you plug in. If not there could be a fault with the Custard Pi 1 or it has not been plugged in properly.

There is a risk of shorting between the pins on the base of the Custard Pi 1 and some of the components of the Raspberry Pi, such as the HDMI connector or capacitor C6. For this reason, the Custard Pi is supplied with a length of double sided sticky pad to act as insulation. If the Custard Pi 1 is bought as a kit of parts for self assembly, then sticky pads are supplied and must be used.

Sticky pads to insulate Custard Pi 1 from Raspberry Pi

2.5 Project 1 - Flashing an LED

Driving LEDs from the Custard Pi 1 is very easy. As there is a current limiting resistor built in (1k on early versions, 220 ohm on later versions). All one has to do is to connect an LED between one of the pins on connector J1 and Gnd. Just make sure that the long leg on the LED is connected to the pin and the short leg is connected to Gnd. In the code below, we assume that the LED is connected to pin 11 of J2, which is one of the general purpose I/O pins.

Custard Pi 1 connected to an LED

```
#sample Python code to flash an led
#www.sf-innovations.co.uk
import RPi.GPIO as GPIO              # import GPIO library
import time                          #import time library
GPIO.setmode(GPIO.BOARD)             #use board pin numbers

GPIO.setup(11, GPIO.OUT)             #setup pin 11 as output

for x in range (0,10):               #repeat for x=0 to 9
    GPIO.output(11, True)            #set pin 11 high
    time.sleep(0.2)                  #wait 0.2 seconds
    GPIO.output(11, False)           #set pin 11 low
    time.sleep(0.2)                  #wait 0.2 seconds

GPIO.cleanup()                       #tidy up GPIO port
import sys                           #exit program
sys.exit()
```

If you would like the LED to flash faster, then change the `time.sleep(0.2)` to a smaller value. For example `time.sleep(0.1)` would make the LED flash twice as fast. Both the time.sleep commands will need to be changed to halve the LED ON time and the LED OFF time.

If you would like the LED to carry on flashing 50 times, instead of just 10, then change the command "`for x in range (0,10):`" to "`for x in range (0,50):`".

The GPIO.setmode command uses the board pin numbers as opposed to the port numbers of the IC used to control the GPIO port. In my experience it is much easier to use the pin numbers as these are clearly identified on the Custard Pi board.

2.6 Reading a Switch

In this mini project, we look at reading a switch and flash the LED only when the switch is pressed. The LED is connected to pin 11 as before. Connect the switch between pin 12 and Gnd. When the switch is pressed, pin 12 will be taken low. The Python code for this is presented below.

Custard Pi 1 connected to a switch and an LED

```
#sample Python code to flash an led when a switch is pressed
#www.sf-innovations.co.uk
import RPi.GPIO as GPIO          # import GPIO library
import time                      #import time library
GPIO.setmode(GPIO.BOARD)         #use board pin numbers
GPIO.setwarnings(False)
GPIO.setup(11, GPIO.OUT)         #setup pin 11 as output
GPIO.setup(12, GPIO.IN, pull_up_down=GPIO.PUD_UP)
                                 #setup pin 12 as input with
pull up
while True:                      #do forever
    while GPIO.input(12)==False: #while switch is pressed
        GPIO.output(11, True)    #set pin 11 high
        time.sleep(0.2)          #wait 0.2 seconds
        GPIO.output(11, False)   #set pin 11 low
```

Raspberry Pi with Custard

```
        time.sleep(0.2)         #wait 0.2 seconds
GPIO.cleanup()                  #tidy up GPIO port
import sys                      #exit program
sys.exit()
```

This code is similar to the previous code but the LED flash is only executed if pin 12 goes low (FALSE) when the switch is pressed. Otherwise the "while True" command keeps the program in an endless loop, waiting for the switch to be pressed.

To exit the program, the user has to press CTRL and C at the same time on the keyboard. Because this exits the program without cleaning up the GPIO interface, we use the command "GPIO.setwarnings(False)" command to stop any warnings from being displayed.

2.7   Electronic Dice

This project uses a 7-segment display and a switch to simulate the roll of a dice. We will use 7 pins from J2 as outputs to drive the 7-segment display and the 8th pin as an input to read the switch. The drawing below shows how to connect up the 7-segment display to the Custard Pi 1.

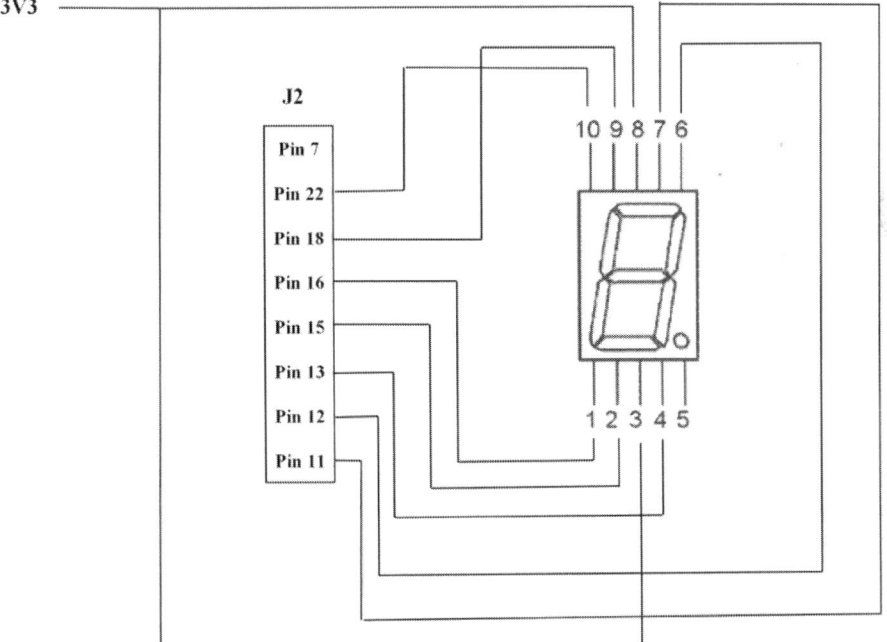

Connecting the 7-segment display to the Custard Pi 1

Connect a switch between pin 7 of connector J2 and Gnd so that pin7 goes low (False) when the switch is pressed.

Raspberry Pi with Custard

Electronic Dice using the Custard Pi 1

The Python code for the electronic Dice is presented below. When the program is started, the 7-segment shows the digit 0. When the switch is pressed, the 7-segment display will randomly display a digit from 1 to 6. This will stay on the display until the switch is pressed again.

```
#!/usr/bin/env python
#sample Python code to display a random digit
#from 1 to 6 when a switch is pressed
#www.sf-innovations.co.uk
import RPi.GPIO as GPIO
import time
import random

GPIO.setwarnings(False)

GPIO.setmode(GPIO.BOARD)

#setup output pins
GPIO.setup(11, GPIO.OUT)
GPIO.setup(12, GPIO.OUT)
GPIO.setup(13, GPIO.OUT)
GPIO.setup(15, GPIO.OUT)
GPIO.setup(16, GPIO.OUT)
GPIO.setup(18, GPIO.OUT)
GPIO.setup(22, GPIO.OUT)
#setup inpt pin with pull up resistor
GPIO.setup(7, GPIO.IN, pull_up_down=GPIO.PUD_UP)

#define 7 segment digits
digitclr=[1,1,1,1,1,1,1]
digit0=[0,0,0,0,0,0,1]
digit1=[1,0,0,1,1,1,1]
digit2=[0,0,1,0,0,1,0]
digit3=[0,0,0,0,1,1,0]
digit4=[1,0,0,1,1,0,0]
digit5=[0,1,0,0,1,0,0]
```

Raspberry Pi with Custard

```python
digit6=[0,1,0,0,0,0,0]

gpin=[11,12,13,15,16,18,22]

#routine to clear and then write to display
def digdisp(digit):
    for x in range (0,7):
        GPIO.output(gpin[x], digitclr[x])
    time.sleep(0.5)
    for x in range (0,7):
        GPIO.output(gpin[x], digit[x])

#wait for switch to be released
def swwait():
    while GPIO.input(7)==False:
        time.sleep(0.1)

#display random digit
def randigit(digit):
    digdisp(digit)
    swwait()

#initialise by clearing display and writing 0
for x in range (0,7):
        GPIO.output(gpin[x], digitclr[x])
digdisp (digit0)

#main routine to read switch and display random digit from 1 to 6
while True:
    if GPIO.input(7)==False:
        rand = random.randint(1,6)
        if rand == 1:
            randigit(digit1)
        if rand == 2:
            randigit(digit2)
        if rand == 3:
            randigit(digit3)
        if rand == 4:
            randigit(digit4)
        if rand == 5:
            randigit(digit5)
        if rand == 6:
            randigit(digit6)

#tidy up
GPIO.cleanup()
import sys
sys.exit()
```

CHAPTER 3   CUSTARD PI 2
General Purpose input/output board for the Raspberry Pi

3.1   Introduction

The Custard Pi 2 board simply plugs into the Raspberry Pi GPIO connector and provides 4 digital outputs, 4 digital inputs, 2 analogue outputs and 2 analogue inputs. At the same time it protects the Raspberry Pi from possible damage from the wrong voltage being accidentally connected to the GPIO pins.

Custard Pi 2

3.2   Circuit Description

When the Custard Pi 2 is plugged into the GPIO, two LEDs come ON, showing that the 5V and 3.3V rail are working correctly. The connections to the GPIO are detailed below.

Pin 1: 3.3V supply
Pin 2: 5V supply
Pin 6: 0V
Pins 7, 16, 18, 22: Digital inputs
Pins 11, 12, 13 and 15: Digital outputs
Pin 19: SPI MOSI
Pin 21: SPI MISO
Pin 23: SPI SCLK
Pin 24: SPI CS0
Pin 26: SPI CS1

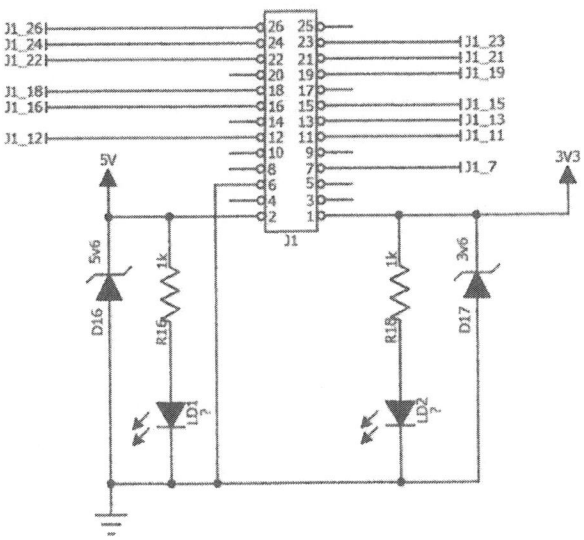

GPIO connections showing power rails

The 3.3V is supplied on pin 1 of the GPIO and the 5V is supplied on pin 2. The 2 LEDs are connected to these pins with a 1k current limiting resistor.

Power pins (J3)

These are brought out to connector J3 on the Custard Pi 2 and have a fuse fitted to each line. This is to prevent the user from drawing too much current from the Raspberry Pi. The fuses are resettable and are both rated at 0.1 Amp (100 m Amp).

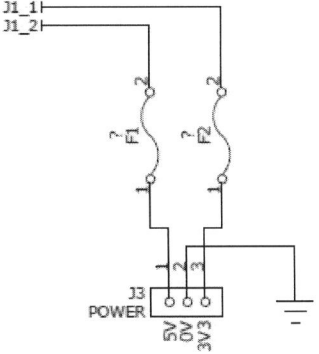

5V and 3.3V pins with fuses

Open Collector Digital Output pins (J2)

The pins marked green above are general purpose digital input output pins.

Pins 11, 12, 13 and 15 are buffered using a ULN2801 IC and brought out to connector J2.

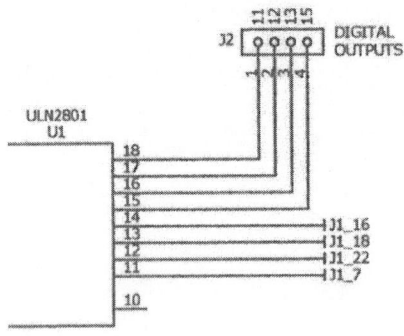

Open Collector Digital outputs

The ULN 2801 has Darlington transistors at the output pins 15, 16, 17 and 18 of U1. This is shown below. What this means is that when pin 11 of the GPIO is taken high, pin 18 of U1 goes low (pin 1 of J2). If a load is connected between a 12V supply and pin 1, then this is switched on. The current capability of the output pins of the U1 is 500mA, and the voltage can be taken right up to 50V.

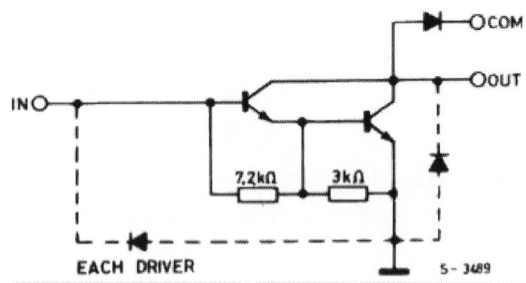

For ULN2801A (each driver for PMOS-CMOS)

Darlington transistor output of ULN2801

Buffered digital inputs

Connector J3 provides 4 buffered input pins. When using these, the GPIO input pins 16, 18, 22 and 7 have to be set with internal pull up resistors on the Raspberry Pi. The buffering is provided by the ULN2801 allowing the pins on J3 to be taken up to any voltage. It is important to use a series resistor to limit the current into the ULN2801 inputs. When J3 pins are taken high, the pins on the GPIO are taken low. When J3 pins are taken low, the pins on the GPIO are taken high - using the internal pull-up resistors.

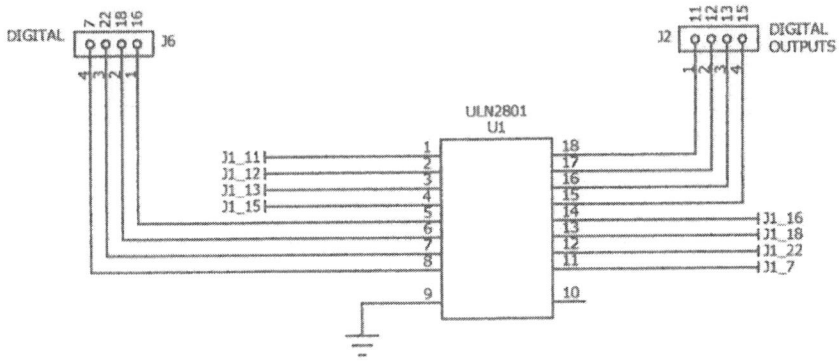

Buffered inputs

12-bit analogue inputs

Two 12-bit input channels are provided by IC U2 which is MCP3202. This is controlled from the GPIO using the SPI bus. This part of the schematic is shown below.

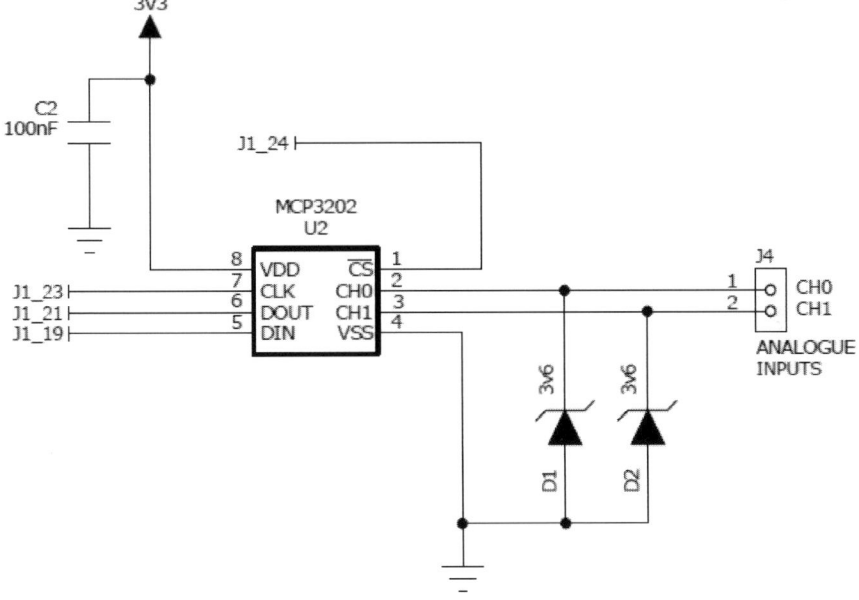

Analogue input channels

The analogue inputs are protected by two 3.6V zener diodes. Full range is from 0 to 3.3V and is represented by 4096 bits. Some example conversions from bits to voltages are shown below.

4096 bits represents 3.3V
2048 bits represents 1.65V

1024 bits represents 0.825V
100 bits represents 0.08V (or 81mV)

12-bit analogue outputs

Two 12-bit output channels are provided by IC U2 which is MCP3202. This is controlled from the GPIO using the SPI bus. This part of the schematic is shown below.

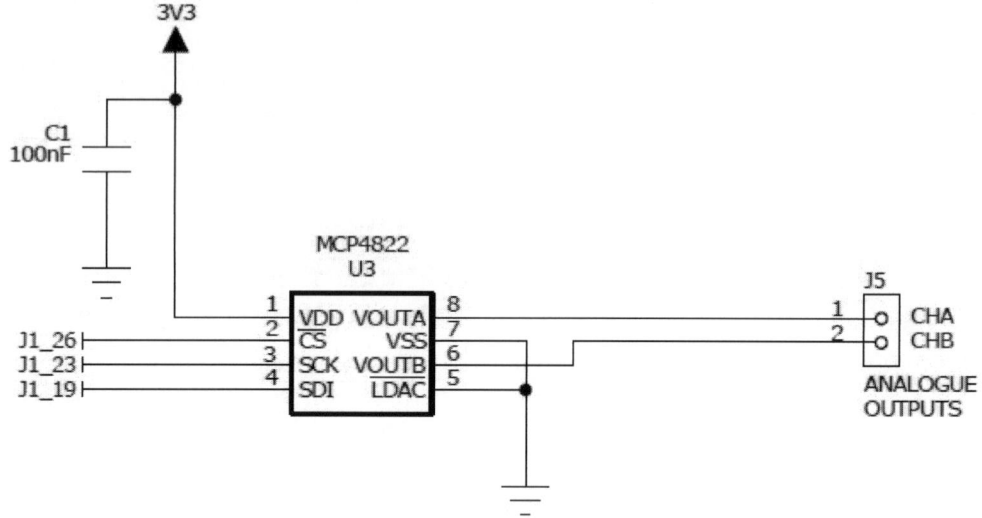

Analogue output channels

The range of output voltages is from 0 to 3.3V.

Notes on the SPI serial bus (J5)

This serial bus is used to interface to a number of external integrated circuits. However it is different from the I2C bus in that it has separate data out and data in lines and the devices are not addressable. There are separate chip enable lines for each integrated circuit. The Raspberry Pi SPI bus is provided with 2 chip enable outputs.

3.3 The Custard Pi 2 Assembly

The positions of the connectors are shown below. These are mini screw terminals into which wires can be quickly connected.

Raspberry Pi with Custard

- 2 analogue inputs
- 2 analogue outputs
- 2 multifuses on 3.3V and 5V rail
- power on leds
- power connector
- over voltage protection
- plug in connection to Raspberry Pi
- 4 digital outputs
- 4 digital inputs

Positions of the connectors

This is a compact assembly that simply plugs into the Raspberry Pi GPIO. This can be done even with the Raspberry Pi is powered. Just make sure that the 2 power LEDs are on as soon as you plug in. If not there could be a fault with the Custard Pi 1 or it has not been plugged in properly.

There is a risk of shorting between the pins on the base of the Custard Pi 2 and some of the components of the Raspberry Pi, such as the HDMI connector or capacitor C6. For this reason, the Custard Pi is supplied with a length of double sided sticky pad to act as insulation. If the Custard Pi 2 is bought as a kit of parts for self assembly, then sticky pads are supplied and must be used.

Sticky pads to insulate Custard Pi 2 from Raspberry Pi

3.4 Sample Python code for testing digital outputs

This test program sets the 4 outputs high and low and repeats 20 times. Connect an led to 3.3V, 5V or any voltage up to 50V through a resistor in series (say 1 k ohm) to limit the current and run the program. Connect the other side of the resistor to Pins 11, 12, 13 or 15 to see it flash.

Note: When GPIO pin is taken high, the output on the Custard Pi goes low (i.e. it is inverted).

```
#1/usr/bin/env python
#Sample Python code to test digital outputs on Custard Pi 2
#www.sf-innovations.co.uk
#This program sets pins 11, 12, 13 & 15 as outputs.
#Sets them all high
#Waits 0.2 seconds
#Sets them all low
#Waits 0.2 seconds
#Repeats 20 timesimport RPi.GPIO as GPIO
import time
GPIO.setmode(GPIO.BOARD)

GPIO.setup(11, GPIO.OUT)
GPIO.setup(12, GPIO.OUT)
GPIO.setup(13, GPIO.OUT)
GPIO.setup(15, GPIO.OUT)

for x in range (0,20):

    GPIO.output(11, True)
    GPIO.output(12, True)
    GPIO.output(13, True)
    GPIO.output(15, True)

    time.sleep(0.2)

    GPIO.output(11, False)
    GPIO.output(12, False)
    GPIO.output(13, False)
    GPIO.output(15, False)

    time.sleep(0.2)
GPIO.cleanup()
import sys
sys.exit()

#TO TEST
#Connect an led to 3.3V, 5V or any voltage up to 50V
#Connect a resistor in series (say 1 kohm) to
#limit the current
```

```
#Connect the other side of the resistor to
#Pins 11, 12, 13 or 15 to see it flash.

#Note: When GPIO pin is taken high, the output on the
#Custard Pi goes low (i.e. it is inverted).
```

3.5 Sample Python code for testing digital inputs

This program reads all 4 digital inputs and prints the results to the screen. Connect high or low signal to the pins 7, 22, 18 and 16 and run the program. For high, do not connect anything to pin (pull up resistor on Raspberry Pi will keep I/P high). For low link pin to 3.3V through a resistor (i.e. these are inverting inputs).

```
#1/usr/bin/env python
#sample python code to test digital inputs on Custard Pi 2
#www.sf-innovations.co.uk
#This program sets up pins 7,22,18 & 16
#As inputs with a pull up resistor
#Scans all 4 inputs
#Prints results to screen
#Waits 1 second
#Repeats 10 times
import RPi.GPIO as GPIO
import time
GPIO.setmode(GPIO.BOARD)

GPIO.setup(7, GPIO.IN, pull_up_down=GPIO.PUD_UP)
GPIO.setup(22, GPIO.IN, pull_up_down=GPIO.PUD_UP)
GPIO.setup(18, GPIO.IN, pull_up_down=GPIO.PUD_UP)
GPIO.setup(16, GPIO.IN, pull_up_down=GPIO.PUD_UP)


for x in range (0,10):
    bit1=GPIO.input(7)
    bit2=GPIO.input(22)
    bit3=GPIO.input(18)
    bit4=GPIO.input(16)
    print bit1, bit2, bit3, bit4
    time.sleep(1)
GPIO.cleanup()
import sys
sys.exit()

#TO TEST
#For high, do not connect anything to pin
#(pull up on Raspberry Pi will keep I/P high)

#Revision 1 (no Custard Pi 2 legend on PCB and 4 diodes on the back
of the PCB)
#For low link pin to 0 V (ie non-inverting inputs)
```

#Revision 2 (Custard Pi 2 legend on PCB and no diodes on the back of the PCB).
#For low link pin to 3.3V (ie inverting inputs)

3.6 Sample Python code for testing analogue inputs

This program reads the voltage on Channel 0 and prints it to the screen. Connect Channel 0 to 3.3V and when you run the program 3.3V should be written to the screen. Connect two equal resistors in series from 3.3V to 0V and connect the mid-point of resistors to Channel 0 input. Voltage printed should be 1.65V

Note: To test Channel 1, set the 3rd bit Low in Word1 in the code below and run the program.

```
#!/usr/bin/env python
#Sample Python code to test analogue inputs on Custard Pi 2
#www.sf-innovations.co.uk
#Program sets pins 24, 23 and 19 as outputs
#Pin 21 as input
#24 - chip enable
#23 - clock
#19 - data out
#21 - data in
#Word1
#1st bit High - start bit
#2nd bit High - two separate channels
#3rd bit High - input on Channel 1
#4th bit High - Most significant bit first
#5th bit High - Clock in null bit
#Chip enable (pin 24) Low
#Clock out word1 bit by bit on pin 19
#Data valid when clock (pin 23) goes from L to H
#Clock in 12 bits on pin 21
#Data valid when clock (pin 23) goes from L to H
#Work out decimal value based on position & value of bit
#X = position of bit
#Bit = 0 or 1
#Value = value of that bit
#anip = running total
#Print x, bit, value, anip
#Chip enable high
#Work out and print voltage


import RPi.GPIO as GPIO
import time
GPIO.setmode(GPIO.BOARD)

GPIO.setup(24, GPIO.OUT)
GPIO.setup(23, GPIO.OUT)
GPIO.setup(19, GPIO.OUT)
GPIO.setup(21, GPIO.IN)
```

Raspberry Pi with Custard

```python
GPIO.output(24, True)
GPIO.output(23, False)
GPIO.output(19, True)

word1= [1, 1, 1, 1, 1]

GPIO.output(24, False)
anip=0

for x in range (0,5):
    GPIO.output(19, word1[x])
    time.sleep(0.01)
    GPIO.output(23, True)
    time.sleep(0.01)
    GPIO.output(23, False)

for x in range (0,12):
    GPIO.output(23,True)
    time.sleep(0.01)
    bit=GPIO.input(21)
    time.sleep(0.01)
    GPIO.output(23,False)
    value=bit*2**(12-x-1)
    anip=anip+value
    print x, bit, value, anip

GPIO.output(24, True)

volt = anip*3.3/4096
print volt

GPIO.cleanup()
import sys
sys.exit()

#TO TEST
#Connect Channel 0 to 3.3V and voltage should be 3.3V
#Connect two equal resistors in series from 3.3V to 0V.
#Connect mid point of resistors to Channel 0 input
#Voltage should be 1.65V
```

3.7 Sample Python code for testing analogue outputs

This program sets Channel B to 2V for 5 seconds and then to 1V for 5 seconds. After repeating this 5 times it stops. Run the program and use a volt-meter on channel B to see the output cycle from 2V to 1V five times.

Note: Change the first bit of Word 1 and Word 2 to test Channel B.

```python
#!/usr/bin/env python
#Sample Python code to test analogue outputs on Custard Pi 2
#www.sf-innovations.co.uk
#Program sets pins 26, 23 and 19 as outputs
#26 - chip enable
#23 - clock
```

```
#19 - data out
#Word1
#1st bit High - writing to Channel B
#2nd bit Low - either state is OK
#3rd bit High - when all bits high, V = 2.048V
#4th bit High - output available
#Last 12 bits all High = 4096 = 2.048V
#Word2
#Last 12 bits =011111111111= 2048 = 1.024 V
#Chip enable (pin 26) Low
#Clock out word1 bit by bit on pin 19
#Data valid when clock (pin 23) goes from L to H
#Chip enable High
#This makes analogue voltage available (word 1)
#Wait 5 seconds
#Chip enable Low
#Clock out word2 bit by bit on pin 19
#Data valid when clock (pin 23) goes from L to H
#Chip enable High
#This makes analogue voltage available (word 2)
#Wait 5 seconds
#Repeat 5 times (using variable count)

import RPi.GPIO as GPIO
import time
GPIO.setmode(GPIO.BOARD)

GPIO.setup(26, GPIO.OUT)
GPIO.setup(23, GPIO.OUT)
GPIO.setup(19, GPIO.OUT)

GPIO.output(26, True)
GPIO.output(23, False)
GPIO.output(19, True)

count = 0
word1= [1, 0, 1, 1, 1, 1, 1, 1, 1, 1, 1, 1, 1, 1, 1, 1]
word2= [1, 0, 1, 1, 0, 1, 1, 1, 1, 1, 1, 1, 1, 1, 1, 1]
while count<5:
    GPIO.output(26, False)

    for x in range (0,16):
        GPIO.output(19, word1[x])
        print word1[x]
        time.sleep(0.01)
        GPIO.output(23, True)
        time.sleep(0.01)
        GPIO.output(23, False)

    GPIO.output(26, True)
    time.sleep(5)
    GPIO.output(26, False)

    for x in range (0,16):
        GPIO.output(19, word2[x])
        print word2[x]
        time.sleep(0.01)
```

```
            GPIO.output(23, True)
            time.sleep(0.01)
            GPIO.output(23, False)

    GPIO.output(26, True)
    print count
    count = count + 1
    time.sleep(5)

GPIO.cleanup()
import sys
sys.exit()

#TO TEST
#Use multi-meter on channel B to see voltage
#cycle from 2V to 1V five times.
```

CHAPTER 4   CUSTARD PI 3
8 analogue input board with stacking connector for the Raspberry Pi

4.1 Introduction

The Custard Pi 3 board simply plugs into the Raspberry Pi GPIO connector and provides users with 8 analogue input channels. At the same time it protects the Raspberry Pi from possible damage from the wrong voltage being accidentally connected to the GPIO.

Custard Pi 3

Note: The stacking connector allows another accessory card to be plugged into the GPIO pins.

4.2 Circuit Description

When the Custard Pi 3 is plugged into the GPIO, an LED comes ON, showing that the 3.3V rail is working correctly.

The diagram below shows the connections to the GPIO. Pin 1 is the 3.3V supply and pin 6 is the 0V connection. The other 4 pins shows are the connections to the SPI bus and are defined as follows.

Pin 19: SPI MOSI - Data into chip
Pin 21: SPI MISO - Data out of chip
Pin 23: SPI SCLK - Clock
Pin 24: SPI CS1 - Chip select (low to select)

Note: Pin 26 (SPI CS2) is not used as we only use one SPI device on this card. It is available for use for communication with another SPI device.

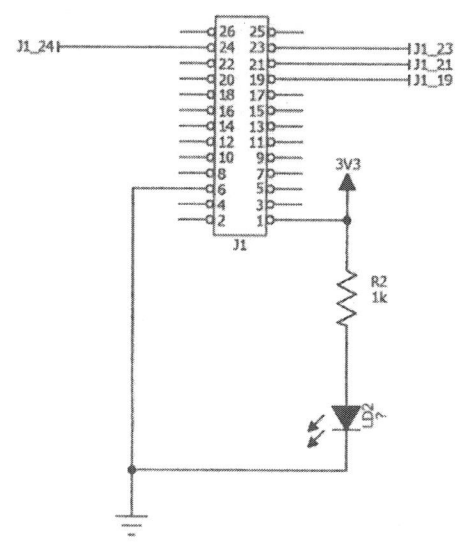

GPIO connections

The connections to the 8 channel Analoge to Digital Convertor (ADC) is shown below. The MCP3208 is a 12-bit convertor (this means it uses 4096 bits to represent voltages from 0 to 2.5V).

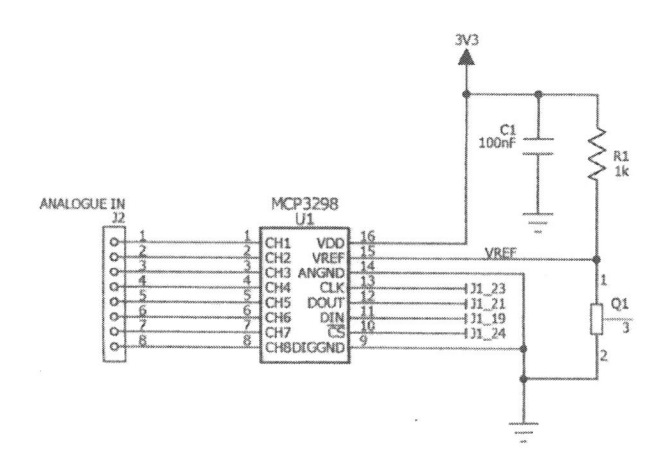

The 8 analogue channels are not voltage limited or protected in any way. It is up to the user to make sure that the analogue voltage presented to the Custard Pi 3 is not larger than 2.5V or more negative than 0V.

Q1 is the external 2.5 precision voltage reference used. Some example conversions from bits to voltages are shown below.

4096 bits represents 2.5V
2048 bits represents 1.25V
1024 bits represents 0.625V

100 bits represents 0.061V (or 61mV)

Some of the signals on the Custard Pi 3 are brought out to a series of via holes and can be used for other purposes, such as providing a 3.3V or 0V supply. This is shown below.

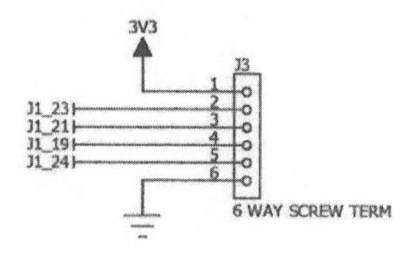

4.3 Sample code for testing analogue inputs

This program reads each analogue channel in turn and prints the result to the screen. Supply suitable analogue voltages between 0 amd 2.5V to the channels and run the program.

Note: we use the terminology Channel 0 to 7. On the PCB they are marked as Channel 1 to 8.

```
#1/usr/bin/env python
#Sample Python program to test 8 analogue inputs on Custard Pi 3
#www.sf-innovations.co.uk

import RPi.GPIO as GPIO
import time
GPIO.setmode(GPIO.BOARD)

GPIO.setup(24, GPIO.OUT)      #pin 24 is chip enable
GPIO.setup(23, GPIO.OUT)      #pin 23 is clock
GPIO.setup(19, GPIO.OUT)      #pin 19 is data out
GPIO.setup(21, GPIO.IN)       #pin 21 is data in

#set pins to default state
GPIO.output(24, True)
GPIO.output(23, False)
GPIO.output(19, True)

#set address channels 0 to 7
#1st bit selects single/differential
#2nd bit channel address
#3rd bit channel address
#4th bit channel address
#5th bit 1 bit delay for data
#6th bit 1st null bit of data
```

Raspberry Pi with Custard

```python
#read each channel in turn
for y in range (0,8):
    if y == 0:
        word= [1 ,1, 0, 0, 0, 1, 1]     #set channel 0
    if y == 1:
        word= [1 ,1, 0, 0, 1, 1, 1]     #set channel 1
    if y == 2:
        word= [1 ,1, 0, 1, 0, 1, 1]     #set channel 2
    if y == 3:
        word= [1 ,1, 0, 1, 1, 1, 1]     #set channel 3
    if y == 4:
        word= [1 ,1, 1, 0, 0, 1, 1]     #set channel 4
    if y == 5:
        word= [1 ,1, 1, 0, 1, 1, 1]     #set channel 5
    if y == 6:
        word= [1 ,1, 1, 1, 0, 1, 1]     #set channel 6
    if y == 7:
        word= [1 ,1, 1, 1, 1, 1, 1]     #set channel 7

    GPIO.output(24, False)   #enable chip
    anip=0   #clear variable

#clock out 7 bits to select channel
    for x in range (0,7):
        GPIO.output(19, word[x])
        time.sleep(0.01)
        GPIO.output(23, True)
        time.sleep(0.01)
        GPIO.output(23, False)

#clock in 11 bits of data
    for x in range (0,12):
        GPIO.output(23,True)       #set clock hi
        time.sleep(0.01)
        bit=GPIO.input(21)         #read input
        time.sleep(0.01)
        GPIO.output(23,False)      #set clock lo
        value=bit*2**(12-x-1)      #work out value of this bit
        anip=anip+value            #add to previous total
#       print x, bit, value, anip

    GPIO.output(24, True)          #disable chip

    volt = anip*2.5/4096           #use ref voltage of 2.5 to work out voltage
    print "voltage ch", y, ("%.2f" %round(volt,2)) #print to screen

GPIO.cleanup()
import sys
sys.exit()

#TO TEST
#Note: we use the terminology Channel 0 to 7. On the PCB they are marked as Channel 1 to 8.
#Connect Channels 0 to 7 to 2.5V and voltage should be 2.5V printed to screen
#Connect two equal resistors in series from 2.5V to 0V.
```

```
#Connect mid point of resistors to Channels 0 to 7 in turn
#Voltage should be 1.25V printed to screen
```

4.4 Weighing Scales project using the Custard Pi 3

Objective: Use the Load Cell pictured below to make a weighing scale.

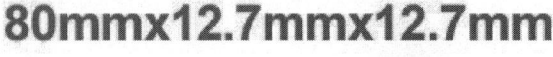

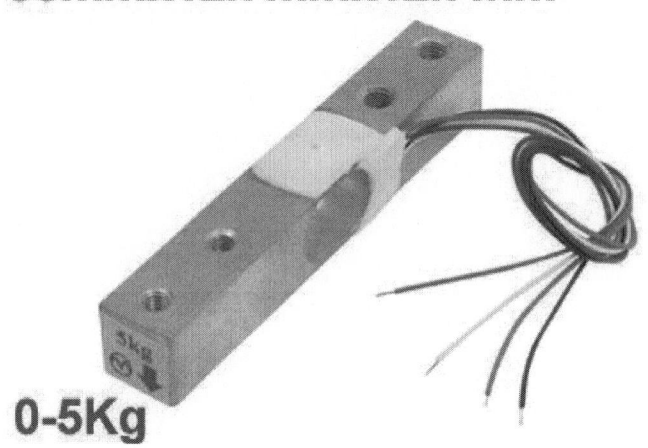

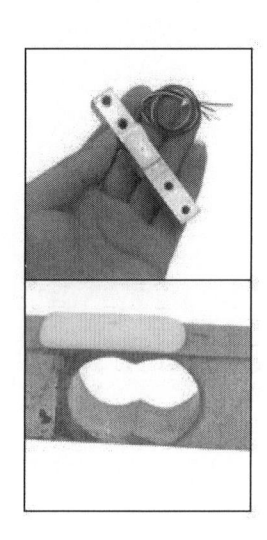

An initial look led to the following findings:

1. This has a 4 wire interface. You need to supply a voltage on the red and black wires. I supplied 12V from a wall charger. The recommended range is from 10 to 15V.

2. With a voltmeter attached to the white and blue wires and a load of approximately 5 kg on the load cell the output was around 5 mV on the white and blue wires.

The input range on the Custard Pi 3 is from 0 to 2.5V. So to get the best accuracy from this load cell, we need an amplification of around 400 using an op-amp.

After this amplification one can feed the output of the op-amp into one of the analogue inputs of the Custard Pi 3 (8 analogue input card) and calibrate the system.

Amplifying the few milli-volts from the load cell up to a usable voltage

We used a simple LM324 for this. The 12V required for the Load Cell was also used to power the LM324. The gain was set to 1000 to get the voltage upto a usable level. The bias voltage was set at around 2.3V. As more and more weight is added, the reading will go down from 2.3V. Full scale deflection on the Custard pi 3 is 2.5V so this fits in nicely.

Here is the circuit diagram.

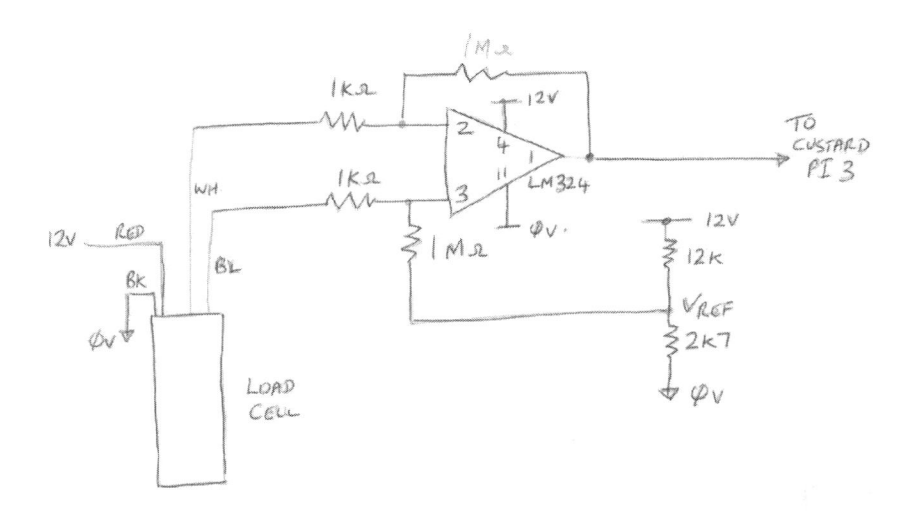

Program to run the scales

A listing of the Python code is given later. There are 2 calibration steps required for the scales. One is to set the zero point - ie the reading when there is nothing on the scales. The second is to determine what the output is for a particular weight. The first time ever the program is run, the user HAS to carry out both these steps. The calibration values are saved in a file on the SD card called "scalesdata.dat". On subsequent running of the program, the user can go straight to weighing as the program uses values saved in the data file.

Challenges ahead

1.   Making a suitable container for items to be weighed. (We just balanced items and depending on how they were placed we got different readings.)

2.   Ensuring that the Load cell is linear - ie the voltage output and the weight have a linear relationship.

Summary

This experimental project was to show how to use an Analogue to Digital Convertor (ADC) to interface to an external transducer, like a Load Cell. It's up-to the reader to develop this further.

Python Code for weighing scales

```python
#1/usr/bin/env python
#Sample Python weighing scales program
#www.sf-innovations.co.uk

import RPi.GPIO as GPIO
import time
import pickle

GPIO.setmode(GPIO.BOARD)
GPIO.setwarnings(False)

#define pins for atod
GPIO.setup(24, GPIO.OUT)     #pin 24 is chip enable
GPIO.setup(23, GPIO.OUT)     #pin 23 is clock
GPIO.setup(19, GPIO.OUT)     #pin 19 is data out
GPIO.setup(21, GPIO.IN)      #pin 21 is data in

#set atod pins to default state
GPIO.output(24, True)
GPIO.output(23, False)
GPIO.output(19, True)

def atod ():
    #This function reads channel 0 and returns a 12 bit figure in "anip"
    word= [1 ,1, 0, 0, 0, 1, 1]
    #set channel 0
    #1st bit selects single/differential
    #2nd bit channel address
    #3rd bit channel address
    #4th bit channel address
    #5th bit 1 bit delay for data
    #6th bit 1st null bit of data
    GPIO.output(24, False)   #enable chip
    anip=0   #clear variable

    #clock out 7 bits to select channel
    for x in range (0,7):
        GPIO.output(19, word[x])
        time.sleep(0.001)
        GPIO.output(23, True)
        time.sleep(0.001)
        GPIO.output(23, False)

    #clock in 11 bits of data
    for x in range (0,12):
        GPIO.output(23,True)     #set clock hi
        time.sleep(0.001)
        bit=GPIO.input(21)       #read input
        time.sleep(0.001)
        GPIO.output(23,False)    #set clock lo
        value=bit*2**(12-x-1)    #work out value of this bit
```

```python
            anip=anip+value          #add to previous total
        #print x, bit, value, anip

    GPIO.output(24, True)        #disable chip

    return anip

#Main Program
while True:
    print"\nThe first time ever the program is run, always select c0 and then cx"
    print "c0 is for calibrating the zero weight. No weight on scale."
    print "cx is for calibrating the scaling factor. Known weight on scale."
    print "w is for continuous weighing. CTRL-C to exit program."

    mode=raw_input("\nEnter w for weighing, c0 for zero weight, cx for known weight ")

    ## value cal0 is saved in file "scaledata". This is the output from the atod for zero weight.
    if mode=="c0":
        analog = atod()
        cal0=analog
        volt = (cal0-analog)*2.5/4096
        #use ref voltage of 2.5 to work out voltage
        print "\natod reading = ",(analog),"zero ref voltage = ",("%.2f" %round(volt,2))
        file = open ("scaledata.dat","wb")
        pickle.dump (cal0,file)
        file.close()

    ## value "scale" is saved to file. Output of atod multiplied by scale gives weight in gms.
    if mode=="cx":
        refweight=int(raw_input("\nenter weight in gms = "))
        print "\nref weight = ", (refweight)
        analog = atod()
        file = open ("scaledata.dat","rb")
        cal0 = pickle.load(file)
        file.close()

        volt = (cal0-analog)*2.5/4096
        #use ref voltage of 2.5 to work out vol
        scale=refweight/volt
        print "\natod reading = ",(analog), "volt = ",("%.2f" %round(volt,2)),"scale factor = ",("%.2f" %round(scale,2))
        file = open ("scaledata.dat", "ab")
        pickle.dump (scale,file)
        file.close()

    ## continuously measures the weight in gms
    if mode =="w":
        while True:
            analog=atod()
            file = open ("scaledata.dat","rb")
```

```
            cal0 = pickle.load(file)
            scale = pickle.load(file)
            file.close()
            volt = (cal0-analog)*2.5/4096         #use ref voltage of 2.5 to work out v
            weight = volt * scale
            #print (analog), "volt",("%.2f" %round(volt,2))
            print "\nweight = ",("%.2f" %round(weight,2)), "gms"
            time.sleep(0.5)
GPIO.cleanup()
import sys
sys.exit()
```

CHAPTER 5   CUSTARD PI 5
Breakout board with protection for 8 digital I/O and stacking connector for the Raspberry Pi GPIO

5.1   Introduction

The Custard Pi 5 board simply plugs into the Raspberry Pi GPIO connector and allows users to quickly connect to all the digital input/output pins. At the same time it protects the Raspberry Pi from possible damage from the wrong voltage being accidentally connected to the GPIO.

Custard Pi 5

Note: The stacking connector allows another accessory card to be plugged into the GPIO pins.

5.2 Circuit Description

When the Custard Pi is plugged into the GPIO, two LEDs come ON, showing that the 5V and 3.3V rail are working correctly. All the other connections to the GPIO are listed below.

Pin 1: 3.3V supply
Pin 2: 5V supply
Pin 6: 0V
Pins 11, 12, 13, 15, 16, 18, 22 and 7: Digital input output pins

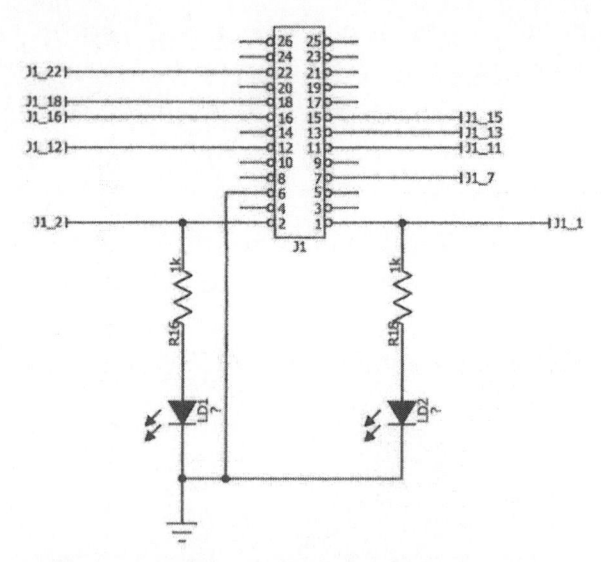

GPIO connections showing power rails

The 3.3V is supplied on pin 1 of the GPIO and the 5V is supplied on pin 2. The 2 LEDs are connected to these pins with a 1k current limiting resistor.

Power pins (J3)

These are brought out to connector J3 on the Custard Pi 5 and have a fuse fitted to each line. This is to prevent the user from drawing too much current from the Raspberry Pi. The fuses are resettable and are both rated at 0.1 Amp (100 m Amp).

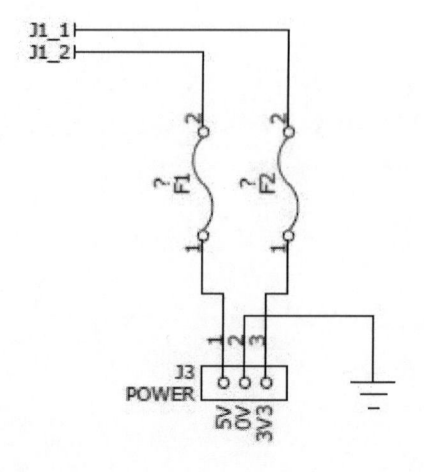

5V and 3.3V pins with fuses

General Purpose Input Output (I/O) pins (J2)

The pins marked green are general purpose digital input output pins. These are pins 11, 12, 13, 15, 16, 18, 22 and 7. They can be set high (to 3.3V) or low (0V) by program control from the Raspberry Pi. The Pi can also read whether these pins are high or low, say to a switch being pressed.

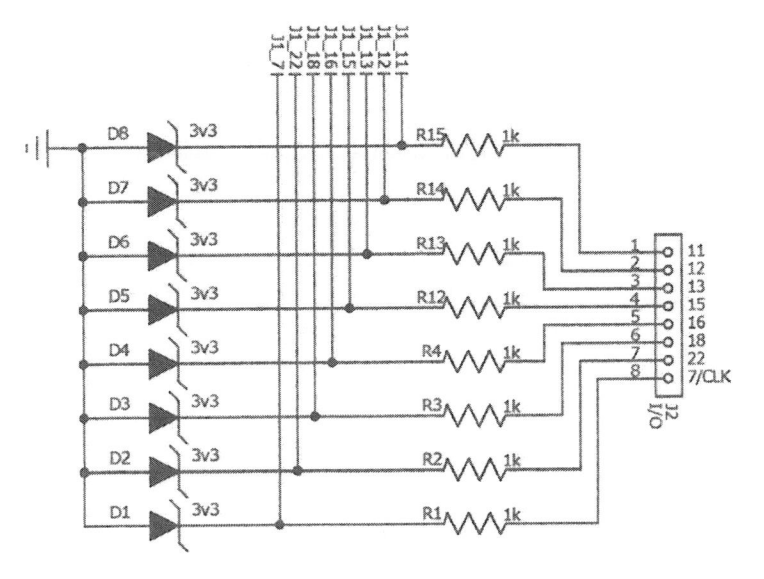

General purpose I/O pins

Each of these pins is protected by a 3.6V (marked as 3.3V on the schematic) zener and a 220 ohm (marked as 1k ohm on schematic) current limiting resistor. The 3.6V zener prevents any voltages in excess of 3.6V from being applied to the pins of the Integrated Circuit on the Raspberry Pi and damaging it. It also protects from negative voltages being applied to the pins. Due to the diode action of the zener voltages on the pins are limited to -0.7V.

5.5 The Custard Pi 5 Assembly

The image below shows the assembled Custard Pi 5 plugged into the Raspberry Pi GPIO.

power connector
3.3 V
0 V
5 V

8 digital I/O on pins 11, 12, 13, 15, 16, 18, 22 and 7

2 multifuses on 3.3V and 5V rail

power on LEDs on 3.3 & 5V

plug in connection to Raspberry Pi using stackable connector

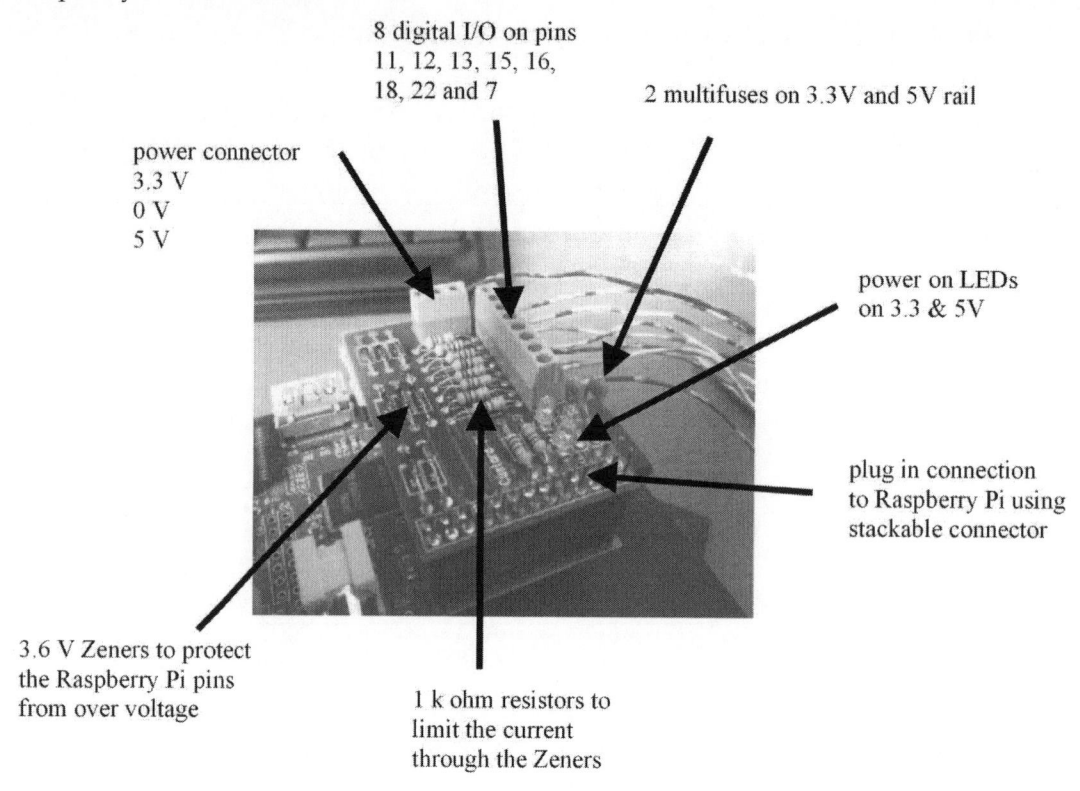

3.6 V Zeners to protect the Raspberry Pi pins from over voltage

1 k ohm resistors to limit the current through the Zeners

This is a compact assembly that simply plugs into the Raspberry Pi GPIO. This can be done even with the Raspberry Pi is powered. Just make sure that the 2 power LEDs are on as soon as you plug in. If not there could be a fault with the Custard Pi 1 or it has not been plugged in properly.

There is a risk of shorting between the pins on the base of the Custard Pi 5 and some of the components of the Raspberry Pi like capacitor C6. For this reason, the Custard Pi is supplied with a length of double sided sticky pad to act as insulation. If the Custard Pi 1 is bought as a kit of parts for self assembly, then sticky pads are supplied and must be used.

Sticky pads to insulate Custard Pi 5 from Raspberry Pi

Note: The Custard Pi 5 is compatible with the Model B and the B+ versions of the Raspberry Pi.

The Custard Pi 1 projects listed in Chapter 2 can easily be adopted for the Custard Pi 5.

CHAPTER 6 CUSTARD PI 6
8 Relay Card for the Raspberry Pi

6.1 Introduction

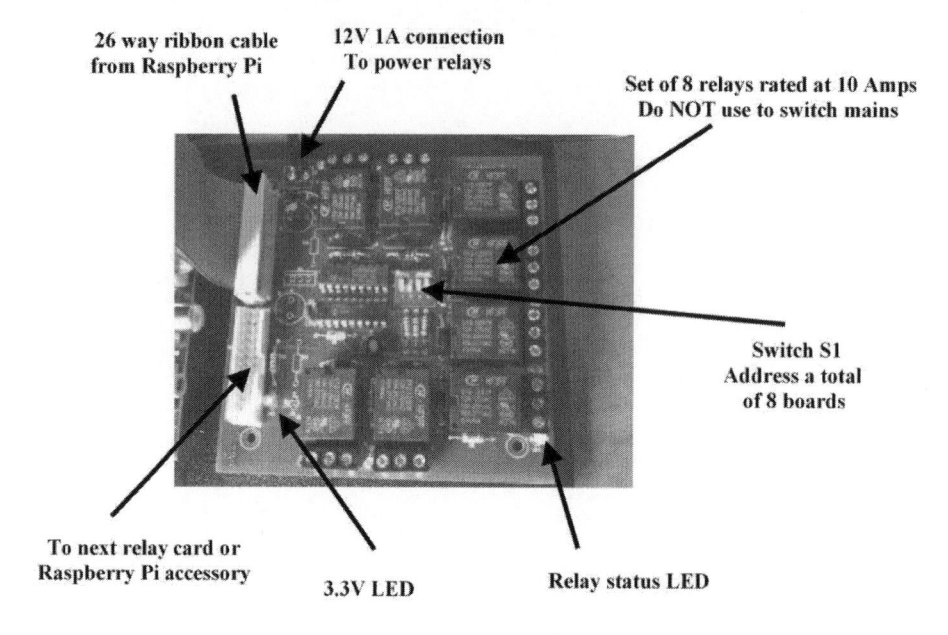

26 way ribbon cable from Raspberry Pi

12V 1A connection To power relays

Set of 8 relays rated at 10 Amps Do NOT use to switch mains

Switch S1 Address a total of 8 boards

To next relay card or Raspberry Pi accessory

3.3V LED

Relay status LED

This chapter describes the Custard Pi 6 circuit which uses the I2C bus to control 8 relays. As the I2C bus is addressable, up to 8 Custard Pi 6s can be daisy chained to provide a total of 64 relays that can be controlled. The only pins on the GPIO that are used are the power pins and the two I2C pins, leaving all the other ports free.

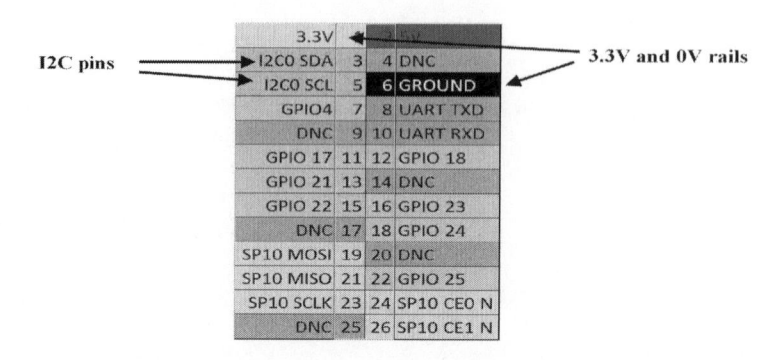

3.3V			
I2C0 SDA	3	4	DNC
I2C0 SCL	5	6	GROUND
GPIO4	7	8	UART TXD
DNC	9	10	UART RXD
GPIO 17	11	12	GPIO 18
GPIO 21	13	14	DNC
GPIO 22	15	16	GPIO 23
DNC	17	18	GPIO 24
SP10 MOSI	19	20	DNC
SP10 MISO	21	22	GPIO 25
SP10 SCLK	23	24	SP10 CE0 N
DNC	25	26	SP10 CE1 N

I2C pins *3.3V and 0V rails*

One 26 way ribbon cable is supplied with the Custard Pi 6 and the configuration of this is shown below.

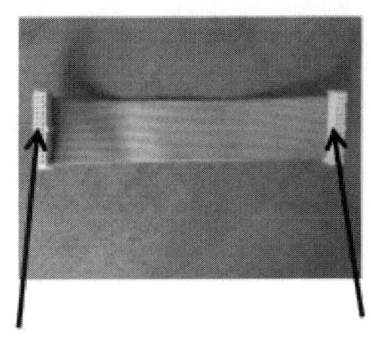

To Raspberry Pi
Notch to the right

To Custard Pi 6
Notch to the right

To maintain compatibility with the Raspberry Pi model B+ which has a 40 pin GPIO connector, an adaptor is supplied with each Custard Pi 6 as shown below.

26 way ribbon cable and adaptor plugged into 40 pin connector

6.2 Circuit description

Connection to the Raspberry Pi

There are two connectors provided on the Custard Pi 6. One is for connection to the Raspberry Pi GPIO and the other is for "daisy chaining" out to either a second Custard Pi 6 PCB assembly, a Custard Pi 3 (8 input analogue card) or some other suitable accessory.

A power on LED is provided on the 3V3 pin to show when the connection is made. The Custard Pi 6 only uses 6 pins on the GPIO and these are listed

Raspberry Pi with Custard

below.

Pin 1, 2 and 0: 3.3V, 5V and 0V
Pin 3: I2C SDA
Pin 5: I2C SCL

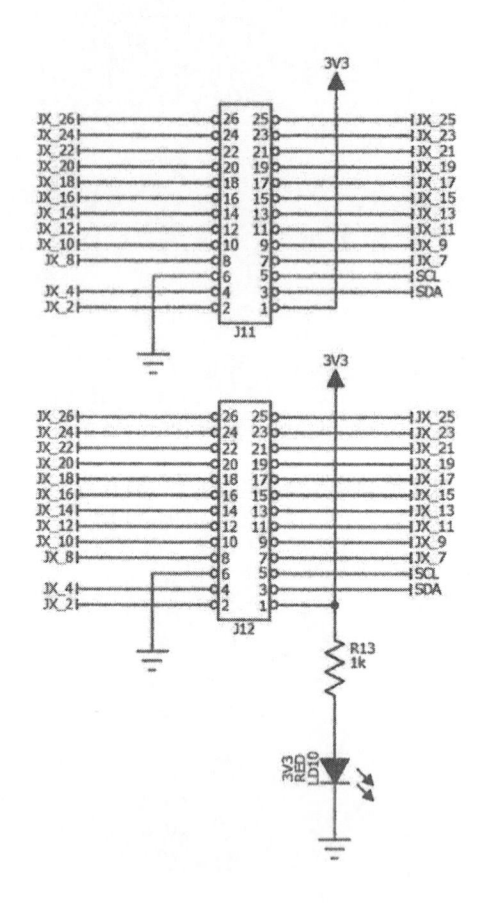

I2C interface

The MCP23008 chip is used to provide 8 ports using the I2C interface and is powered from the 3.3V rail from the GPIO bus. The benefit of using the I2C bus for this is that other than the SDA and SCL pins, the other pins on the GPIO are available to the user when the Custard Pi 6 is in use.

Raspberry Pi with Custard

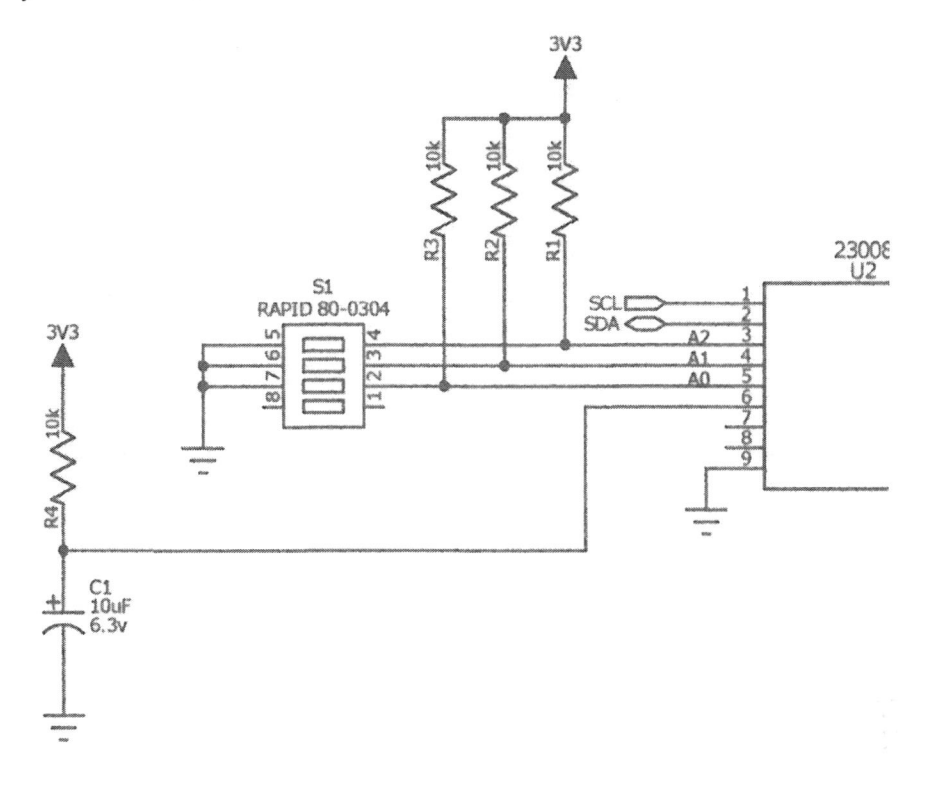

For example, it is possible to drive 8 Custard Pi 6's from one Raspberry Pi and control 64 relays using the I2C bus.

S1 pos 2	S1 pos 3	S1 pos 4	Address
ON	ON	ON	add0
OFF	ON	ON	add1
ON	OFF	ON	add2
OFF	OFF	ON	add3
ON	ON	OFF	add4
OFF	ON	OFF	add5
ON	OFF	OFF	add6
OFF	OFF	OFF	add7

Position 1 is not used.

The 4 way DIL switch allows the user to select the I2C address. This is set as shown below.

Note: add0 to add7 refers to the addresses set in the functions supplied by SF Innovations. If writing your own I2C routines please note that these map as follows.

add 0   0x20

add 1   0x21
add 2   0x22
add 3   0x23
add 4   0x24
add 5   0x25
add 6   0x26
add 7   0x27

Relay Driver

The output from the MCP23008 is not powerful enough to drive the relays directly. A ULN2801 is used as a relay driver. This has open collector Darlington outputs that can sink up to 500mA.

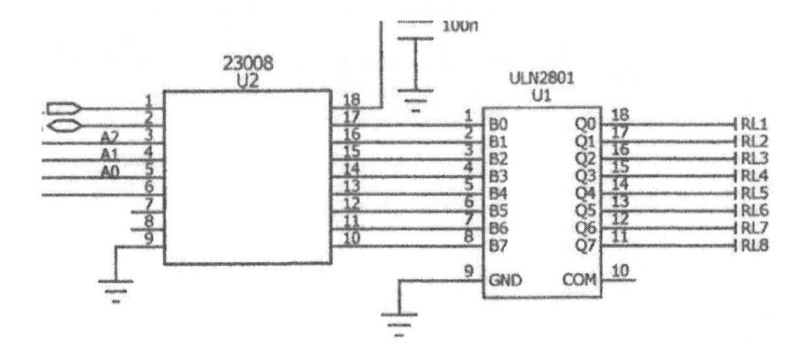

Relay circuit

The standard Custard Pi 6 uses 12V single pole changeover relays. When a relay is switched on an LED also comes on to confirm this. A diode is provided across each relay coil to prevent high voltages being generated when the relay is switched off.

Raspberry Pi with Custard

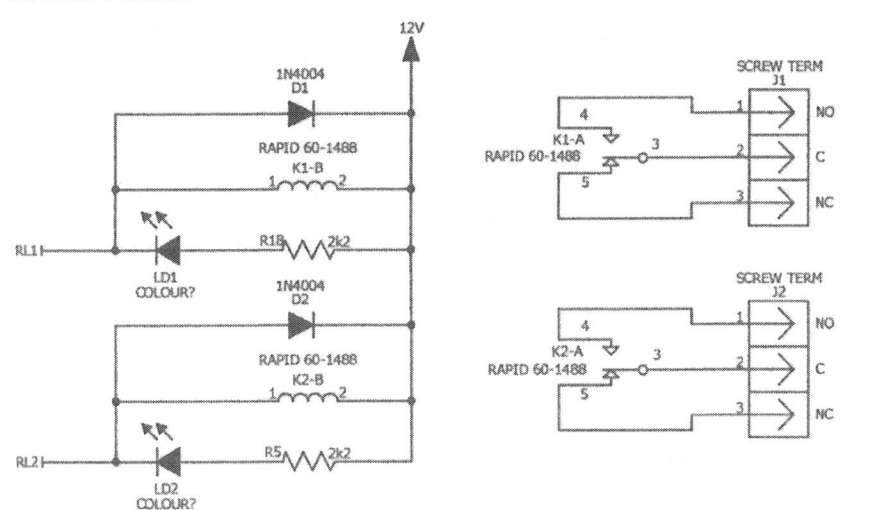

Power supply

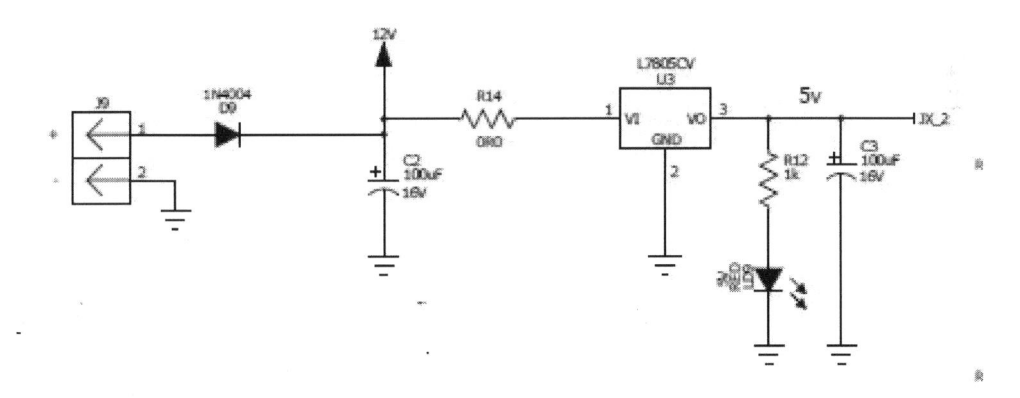

The early version of Custard Pi 6 uses a 12V supply for the relays and the LEDs. Diode D9 is to protect the components if the user happens to connect the supply in the wrong way round. In the 12V version of Custard Pi 6 circuit, components R14, U3, R12, C3 and LD9 are omitted.

Note: When using the 12V version of the Custard Pi 6, a separate 12V power supply needs to be connected to power the relays.

There is now a 5V version of the Custard Pi 6. This means that the Raspberry Pi and the Custard Pi 6 can be powered from the same 5V supply.

Note: One can use a power supply with a micro USB connected to the raspberry Pi to power the 5V version of 8 relay card as well.

59

Raspberry Pi with Custard

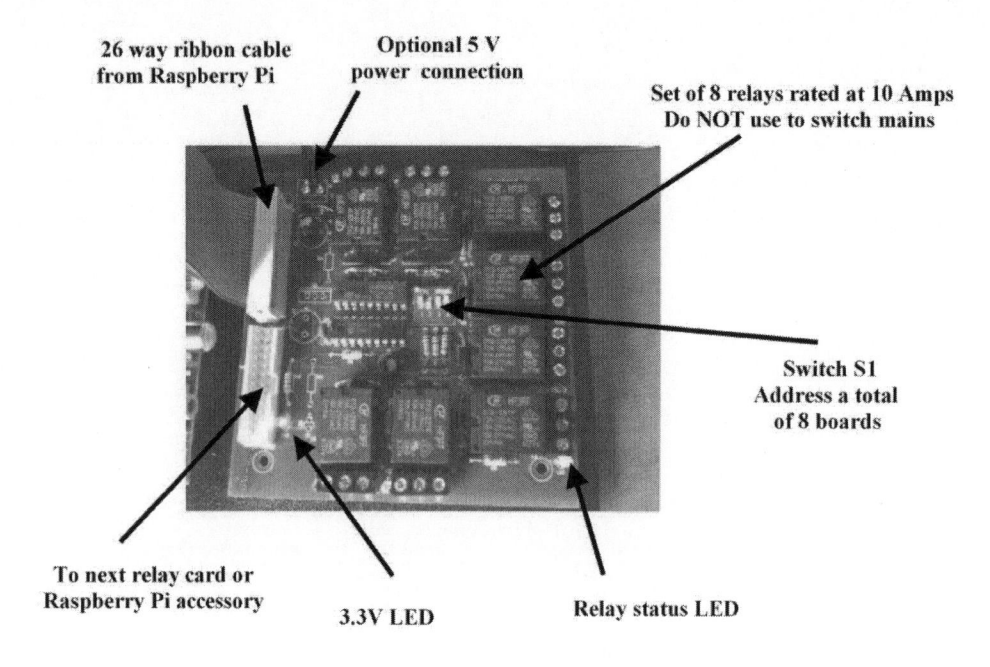

- 26 way ribbon cable from Raspberry Pi
- Optional 5 V power connection
- Set of 8 relays rated at 10 Amps Do NOT use to switch mains
- Switch S1 Address a total of 8 boards
- To next relay card or Raspberry Pi accessory
- 3.3V LED
- Relay status LED

The modifications for the 5V version are shown here.

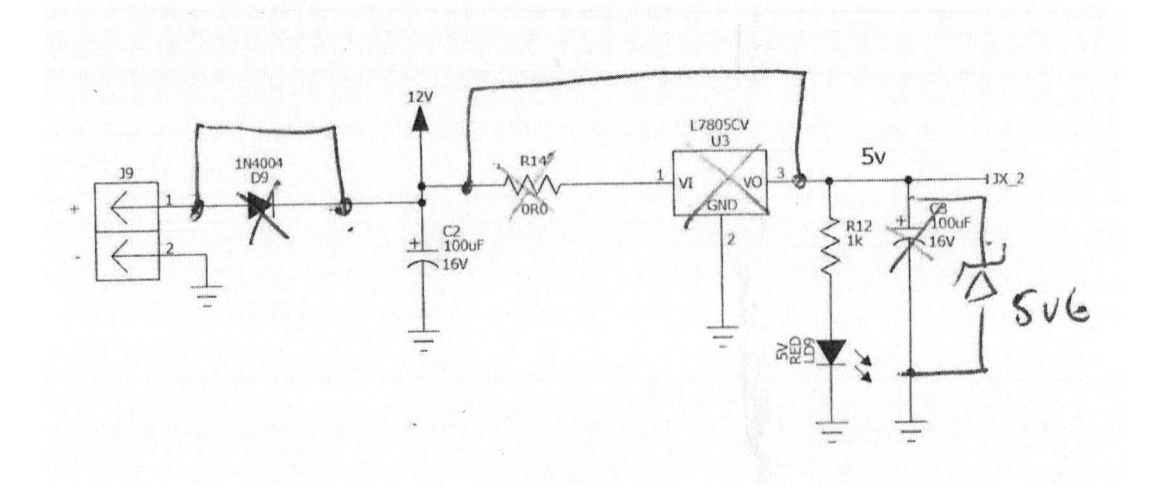

Changes required to the Custard Pi 6 to power it from a single 5V 1A supply. (You can also buy a ready built 5V Custard Pi 6).

1. Fit 5V relays instead of 12V relays.
2. Replace diode D9 with a short circuit link.
3. Make a connection from pin 1 of U3 to pin 3 of U3. (Do not fit U3)
4. Fit R12, and LD9.
5. Fit a 5v6 zener in the position provided for C3. This will prevent the voltage from going higher than 5.6V.

Raspberry Pi with Custard

Please do NOT supply the Raspberry Pi from the micro USB socket. Connect a 5V 1 Amp supply to the Custard Pi 6 PCB using connector J9.

Note: There is no reverse voltage protection so please make sure that the 5V supply is connected the right way round.

Connect the Raspberry pi to the Custard Pi 6 using the ribbon cable. When the Custard Pi 6 is powered using the 5V 1Amp supply, the Raspberry Pi will be powered. It will in turn generate the 3,3V supply and feed this back the Custard Pi 6 PCB assembly. Both the 5V and 3.3V LED will light up when this is working correctly.

6.4 Schematic

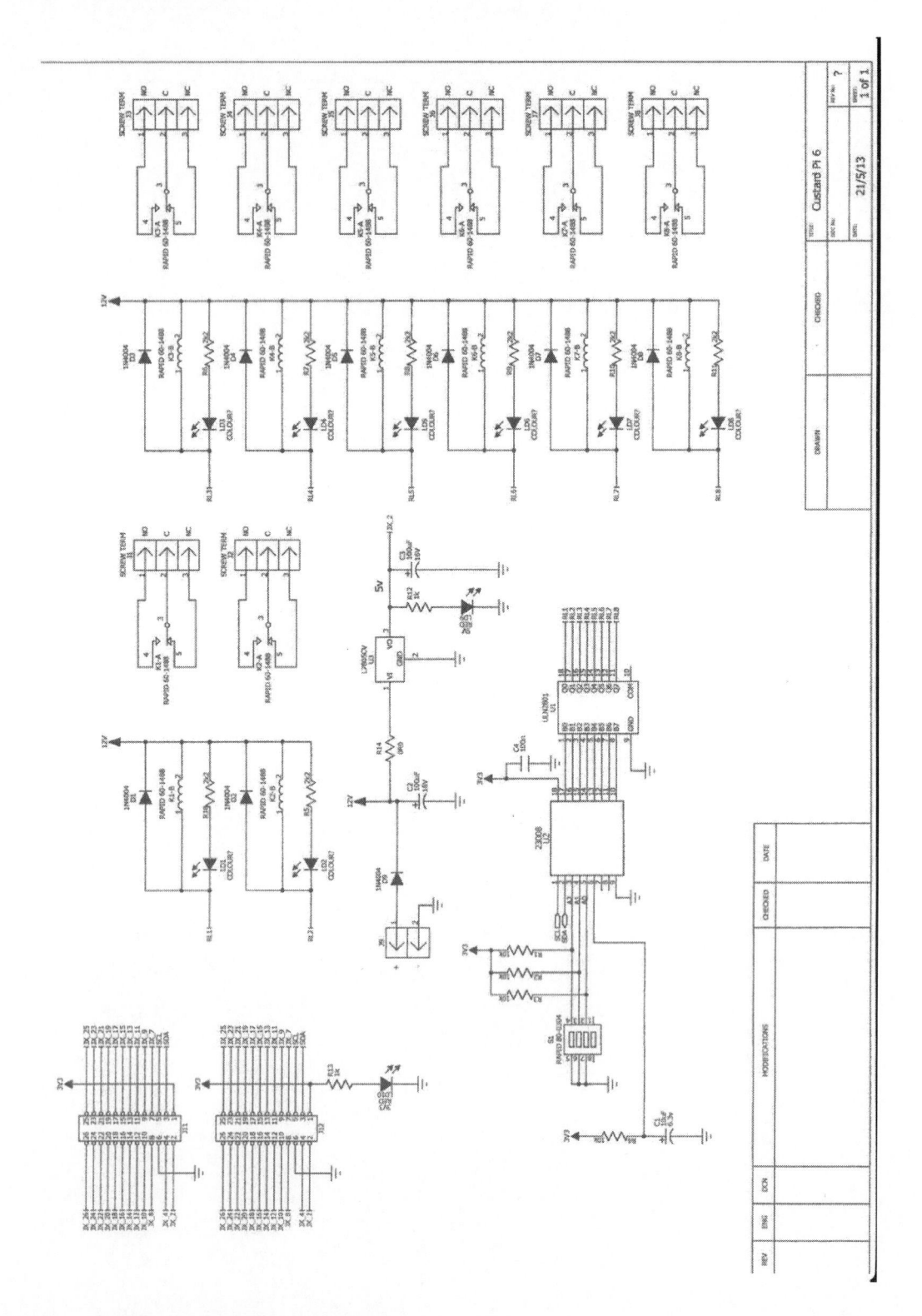

6.5 Setting up the I2C Bus

By default, the I2C bus routines are turned off in the operating system. The following steps need to be followed to enable these.

Step 1

At the command prompt type:

```
sudo nano /etc/modules
```

This uses the nano editor to make some changes to the modules file. Add the following two lines to this file

i2c-bcm2708
i2c-dev

Then save and exit the file using CTRL-x and Y.

Step 2

Make sure that you have the I2C utilities installed by executing the following two commands. The Pi will need to be connected to the Internet for this.

```
sudo apt-get install python-smbus
sudo apt-get install i2c-tools
```

If you get a 404 error do an update first:

```
sudo apt-get update
```

Note : The installation could take a few minutes to do, depend on how busy the server is.

Now add a new user to the i2c group:
```
sudo adduser pi i2c
```

Step 3

On the Raspberry Pi, the I2C and the SPI buses are usually disabled. This is done in the /etc/modprobe.d/raspi-blacklist.conf file.

If this file is not present then there is nothing to be done. Otherwise edit the file by typing the following at the command prompt.

```
sudo nano /etc/modprobe.d/raspi-blacklist.conf
```

If the I2C and the SPI is blacklisted, you will see the following commands.

blacklist spi-bcm2708
blacklist i2c-bcm2708
Insert a # in front of these to comment them out.

Then save and exit the file using CTRL-x and Y.

After editing the file, you will need to reboot for the changes to take effect.

Step 4

Now we need to test if the I2C bus is working correctly.
Connect up the Custard Pi 6 board (or any other I2C bus device) and run the following command.

`sudo i2cdetect -y 1` (for Rev 2 boards which uses port 1)

Or

`sudo i2cdetect -y 0` (for Rev 1 boards which uses port 0)

If everything is OK, then the I2C address of the device will be shown as on the following slide. This shows two devices with address 40 and 70 in hexadecimal code.

Now that we have set up the I2C serial bus routines, we are ready to drive the relays on the Custard Pi 6.

6.6 Driver routine for the Custard Pi 6

We use the function cpi6x.py to control the Custard Pi 6. This is listed below. This function provides the user with the following features.

Addressing a particular Custard Pi 6 card

The user can use add0 through to add7 to directly control up to 8 Custard Pi 6 cards and therefore control a total of 64 relays. Earlier in this chapter we showed how to set up a CPi6 card to a particular address.

Switching a relay ON

This is done by using the 'setbit' command along with the 'ONrelay' parameter. For example, to switch on relay 7 on a Custard Pi6 set to add7, we will use the command 'cpi6x.setbit(board1, cpi6x.ONrelay7)'. The program using this command needs to set board1=cpi6x.add7 to specify a particular Custard Pi 6 board.

Switching a relay OFF

This is done in a similar way with the commands 'clrbit' and 'OFFrelay' being used instead of the 'setbit' and 'ONrelay' commands.

Switching all relays OFF

When initialising the Custard Pi 6 card the command 'alloff' is available to switch all relays OFF.

Initialising the CPi6 card

The command 'setasoutput' is used at the start of any program using this function to initialise the I2C chip.

```
#1/usr/bin/env python
import time
import smbus

#*********************************************
# Custard Pi 6 resources v2.0 9th Sept 2013

#I2C addresses
#use switch S1 on Custard Pi 6 to set the address
```

```python
add0= 0x20
add1= 0x21
add2= 0x22
add3= 0x23
add4= 0x24
add5= 0x25
add6= 0x26
add7= 0x27

bus=smbus.SMBus(1)

#set IODIR register
iodir= 0x00
#set default to all off
allout= 0x00
#set GPIO register
gpio= 0x09
#set output latch
olat=0x0A

#set relay ON
ONrelay0= 0x01
ONrelay1= 0x02
ONrelay2= 0x04
ONrelay3= 0x08
ONrelay4= 0x10
ONrelay5= 0x20
ONrelay6= 0x40
ONrelay7= 0x80

#set relay OFF
OFFrelay0= 0xFE
OFFrelay1= 0xFD
OFFrelay2= 0xFB
OFFrelay3= 0xF7
OFFrelay4= 0xEF
OFFrelay5= 0xDF
OFFrelay6= 0xBF
OFFrelay7= 0x7F

def setbit(address, byte):
    #sets selected port pin
    outstatus = bus.read_byte_data(address, olat) | byte
    bus.write_byte_data(address, gpio, outstatus)

def clrbit(address, byte):
    #clears selected port pin
    outstatus = bus.read_byte_data(address, olat) & byte
    bus.write_byte_data (address, gpio, outstatus)

def setasoutput(address):
    #set all 8 bits as outputs
    bus.write_byte_data(address, iodir, allout)

def alloff(address):
    #clear all relays
```

```
    bus.write_byte_data (address, gpio, 0x00)
```

#***

6.6 Testing out the Custard Pi 6

This program sets relays 0,1&7 on board 1 ON and then turns them all OFF again. This is done continuously until the program is aborted by a CTRL C. When testing, please make sure that switch S1 is set correctly for add1. The program has the line 'import cpi6x' in order to use the commands available in that function.

```
#1/usr/bin/env python
import RPi.GPIO as GPIO
import time
import cpi6x
GPIO.setmode(GPIO.BOARD)
#start program
board1=cpi6x.add1
cpi6x.setasoutput(board1)              #all pins are set as outputs
while True:
    cpi6x.setbit(board1, cpi6x.ONrelay0)        #set relay 0 on board 1 ON
    cpi6x.setbit(board1, cpi6x.ONrelay1)
    cpi6x.setbit(board1, cpi6x.ONrelay7)
    cpi6x.clrbit(board1, cpi6x.OFFrelay0)       #set relay 0 on board 1 OFF
    cpi6x.clrbit(board1, cpi6x.OFFrelay1)
    cpi6x.clrbit(board1, cpi6x.OFFrelay7)
GPIO.cleanup()
import sys
sys.exit()
```

CHAPTER 7   CUSTARD PI 7
Industrial Control Interface for the Raspberry Pi

7.1   General

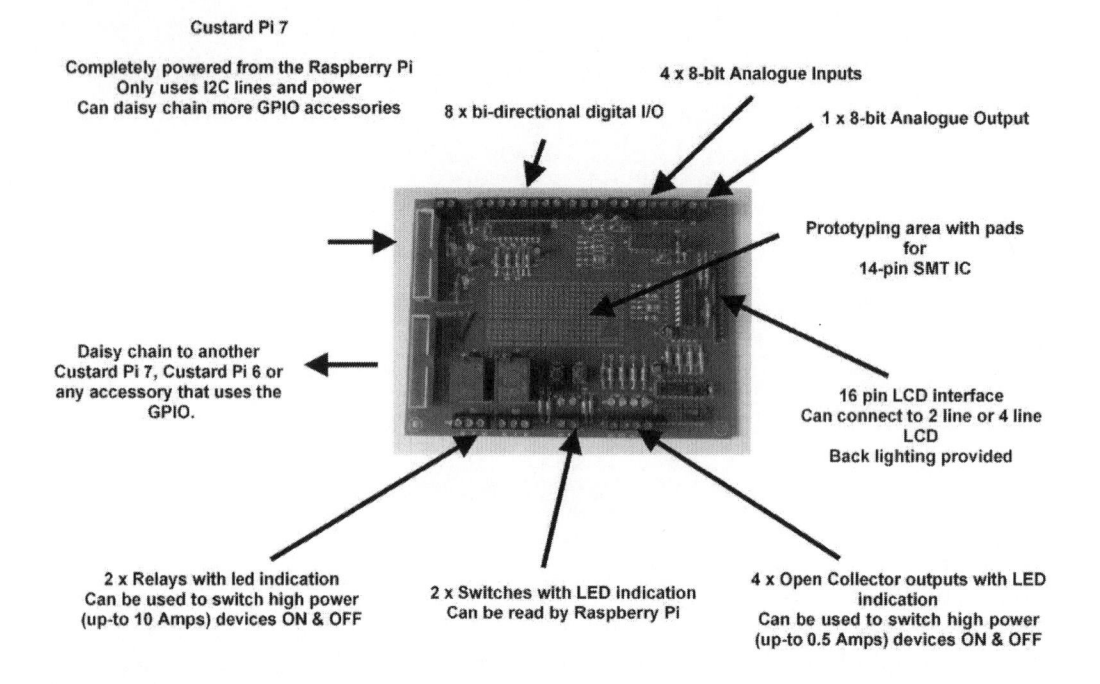

Custard Pi 7

- Completely powered from the Raspberry Pi
- Only uses I2C lines and power
- Can daisy chain more GPIO accessories
- 8 x bi-directional digital I/O
- 4 x 8-bit Analogue Inputs
- 1 x 8-bit Analogue Output
- Prototyping area with pads for 14-pin SMT IC
- Daisy chain to another Custard Pi 7, Custard Pi 6 or any accessory that uses the GPIO.
- 16 pin LCD interface. Can connect to 2 line or 4 line LCD. Back lighting provided
- 2 x Relays with led indication. Can be used to switch high power (up-to 10 Amps) devices ON & OFF
- 2 x Switches with LED indication. Can be read by Raspberry Pi
- 4 x Open Collector outputs with LED indication. Can be used to switch high power (up-to 0.5 Amps) devices ON & OFF

This chapter describes the Custard Pi 7 board, its functions and how to get the most out of it. The drawing above summarises all the features of the CPi7 card.

Like the Custard Pi 6, the 7 is supplied with a ribbon cable for connection to the Raspberry Pi GPIO. The way to connect it is shown below.

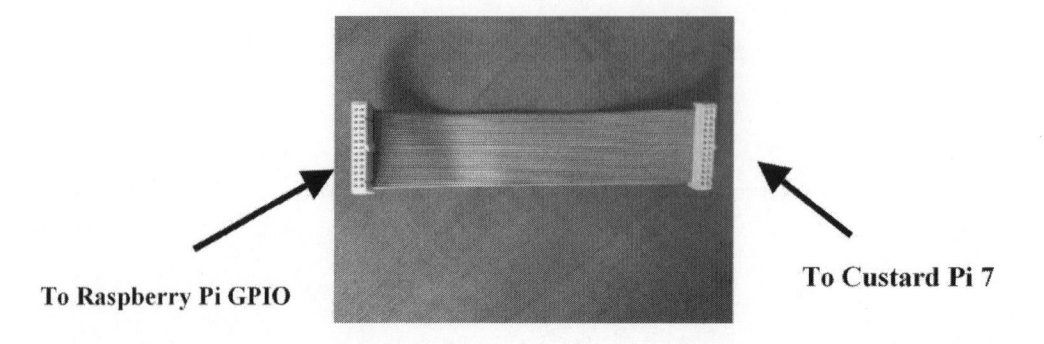

To Raspberry Pi GPIO

To Custard Pi 7

To maintain compatibility with the Raspberry Pi model B+ which has a 40 pin GPIO connector, an adaptor is supplied with each Custard Pi 7.

Raspberry Pi with Custard

26 way ribbon cable and adaptor plugged into 40 pin connector

7.2 Circuit Description

Connection to the Raspberry Pi and Power Supply circuits

There are two connectors provided on the Custard Pi 7. One is for connection to the Raspberry Pi GPIO and the other is for "daisy chaining" out to either a second Custard Pi 7 PCB assembly, a Custard Pi 3 (8 input analogue card) or some other suitable accessory.

As soon as the ribbon cable is plugged in (make sure that the notch is to the right both on the Raspberry Pi and the Custard Pi 7) The 2 LEDs (LD10 & LD11) cone on to show that the 3.3V and 5 V are present.

The Custard Pi 7 only uses 6 pins on the GPIO and these are listed below.

Pin 1, 2 and 0: 3.3V, 5V and 0V
Pin 3: I2C SDA
Pin 5: I2C SCL

The majority of the circuits on the Custard Pi 7 board is powered from the 3.3V supplied from the Raspberry Pi GPIO. This 3.3V is also available on J5 - this output is protected by a 100mA fuse.

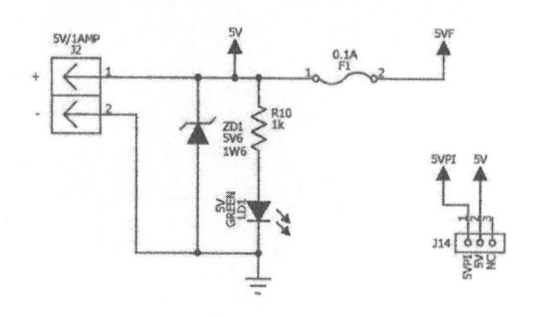

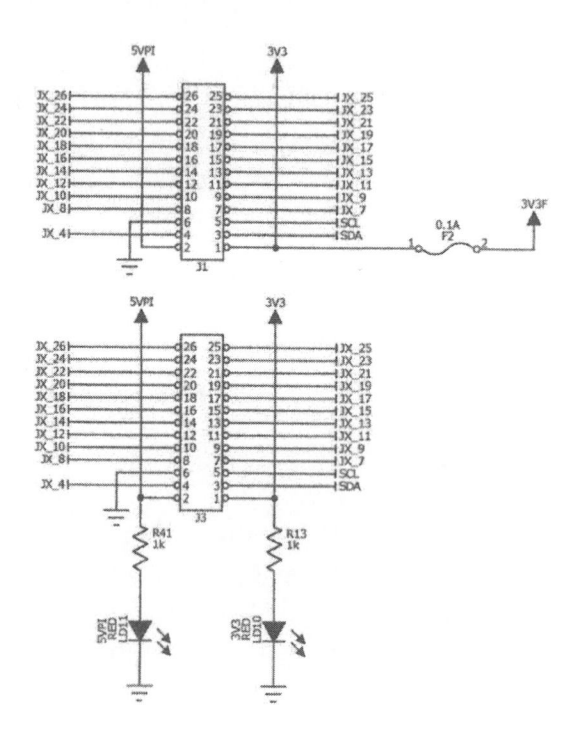

The 5V is used by the 2 relays and the LCD drive circuits. This 5V can be supplied by the GPIO port of the Raspberry Pi or by using an external 5V PSU connected to J2. The user can decide which one to use by making a selection using the jumper on J14. The 5V is also available on connector J5 and is protected by a 100mA fuse.

If the user is planning to power the CPi7 from a single supply (either the RPi supply with a micro USB connector or a 5V supply connected to the CPi7) then please connect the jumper so that it links the 5VPi and 5V pins together. This uses the single supply for the whole board.

If two separate power supplies are being used (one connected to the RPi and another connected to the CPi7) then the jumper is 'parked' on the right hand side and the two supplies are kept separate.

Note: Please make sure that the 5V supply is connected the right way round to connector J2.

MCP23008 I2C chip (U2) providing 6 outputs and 2 inputs

The MCP23008 chip is used to provide 8 ports using the I2C interface and is powered from the 3.3V rail from the GPIO bus. The benefit of using the I2C bus for this is that other than the SDA and SCL pins, the other pins on the GPIO are available to the user when the Custard Pi 7 is in use.

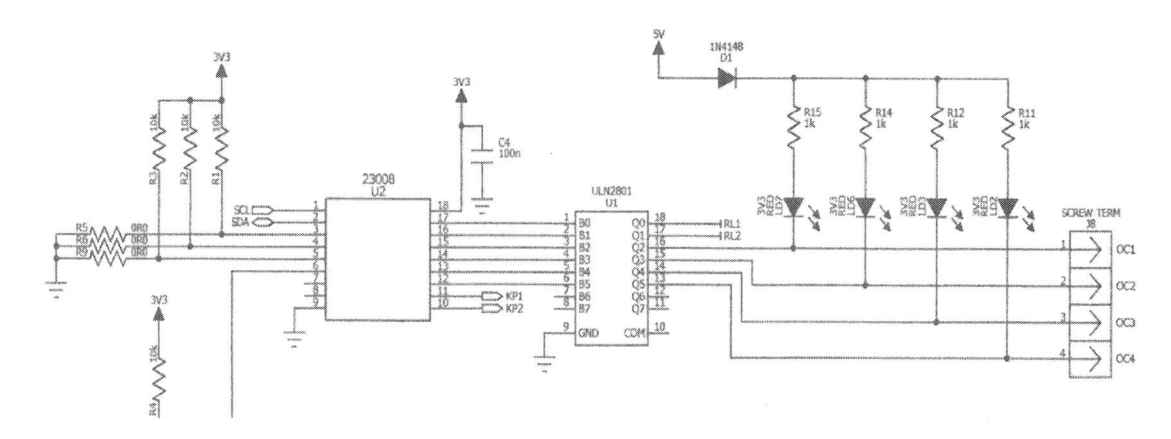

This IC (U2) drives 2 relays and 4 open collector outputs using driver IC ULN2801. It also reads the status of the 2 Keypads built on the board.

The three address pins are pulled high, as R5, R6 & R9 are not fitted. Hence the default I2C address of U2 is 0x27. If one is using more than 1 CPi7 on a system there will be an I2C address conflict. The information below shows how this address can be changed. 'LINK' refers to a 0 ohm link soldered into the position R5, R6 or R9.

R5 fitted?	R6 fitted?	R9 fitted?	I2C address
NO	NO	NO	0X27
NO	NO	LINK	0x26
NO	LINK	NO	0x25
NO	LINK	LINK	0x24
LINK	NO	NO	0x23
LINK	NO	LINK	0x22
LINK	LINK	NO	0x21
LINK	LINK	LINK	0x20

Relay Driver (U1)

The output from the MCP23008 is not powerful enough to drive the relays directly. A ULN2801 is used as a relay driver. This has open collector Darlington outputs that can sink up to 500mA.

Relay outputs

The 2 relays are 6V ones and supplied at 5V either from the Raspberry Pi or an external PSU as outlined before. There are also 2 LEDs that come on when the relays are activated. The 2 connectors, J6 and J7 provide screw terminal access to the contacts of the relays.

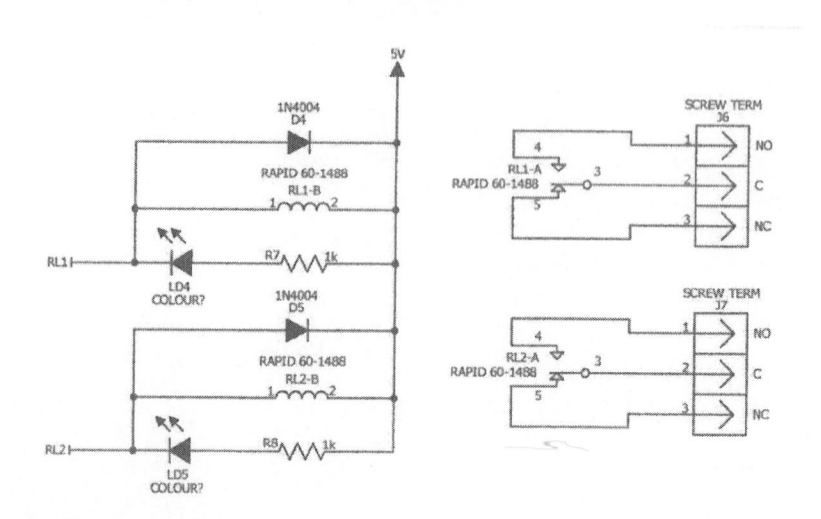

Open Collector outputs

The 4 open collector outputs can be used to switch 500mA each. So for example they can be used to switch relays as well. Their operation is indicated by red LED on each line.

Keypad inputs

There are 2 keys built onto the Custard Pi 7 board. The operation of these can be read by U2 (MCP23008) and confirmed by the LEDs. These inputs are also brought out to screw terminal J9. When the key is pressed the input (KP1 or KP2) goes LOW. They are normally kept HIGH by pull-up resistors R17 and R19.

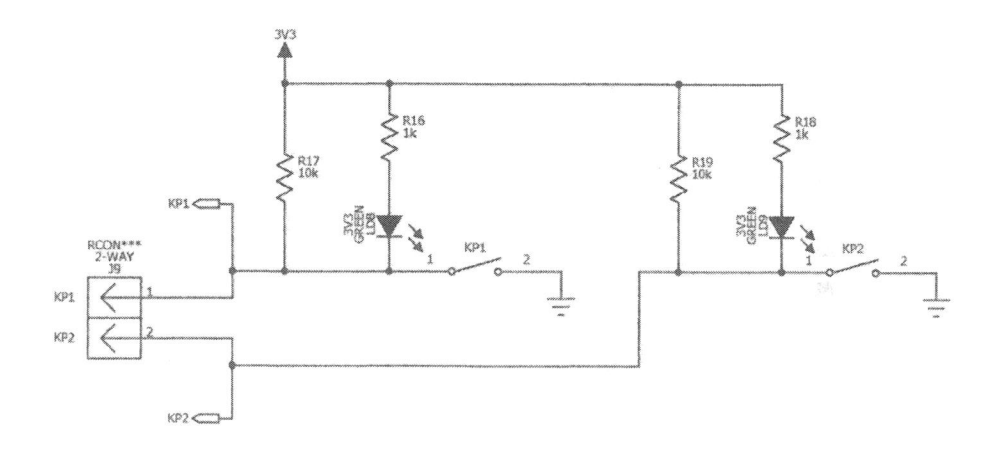

MCP23008 (U3) providing 8 general purpose I/O

This second I2C chip provides 8 general purpose I/O ports that are available on screw terminal J10. The 8 pins can be set as all outputs, all inputs or a mixture.

Resistors R24 and R25 are not fitted while R26 is a 0 ohm link thus setting the I2C address as 0x26. The information below shows how to change the I2C address to avoid addressing conflicts.

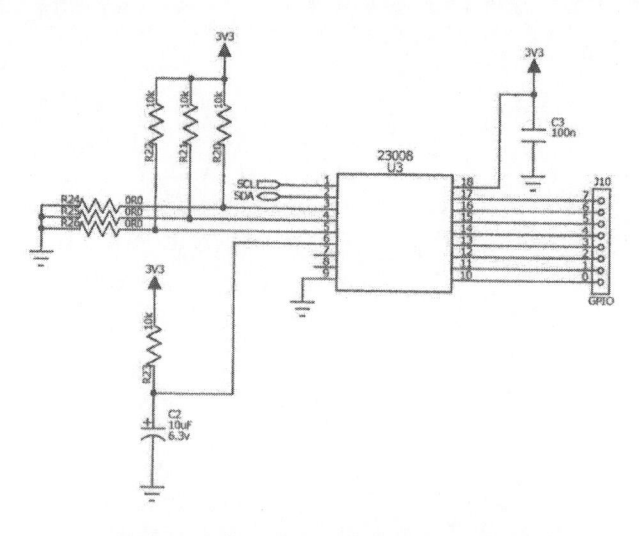

Circuit for the 8 input outputs

R24 fitted?	R25 fitted?	R26 fitted?	I2C address
NO	NO	NO	0X27
NO	NO	LINK	0x26
NO	LINK	NO	0x25
NO	LINK	LINK	0x24
LINK	NO	NO	0x23
LINK	NO	LINK	0x22
LINK	LINK	NO	0x21
LINK	LINK	LINK	0x20

How to change I2C address for U3

Analogue I/O

The chip PCF 8591 provides 4 analogue inputs and 1 analogue output. This also is an I2C bus driven device. The Analogue to Digital Convertor and the Digital to Analogue Converter use 8 bits and a 2.5V external reference. This means that a 2.5V analogue signal will provide a reading of 255 when converted to a digital value.

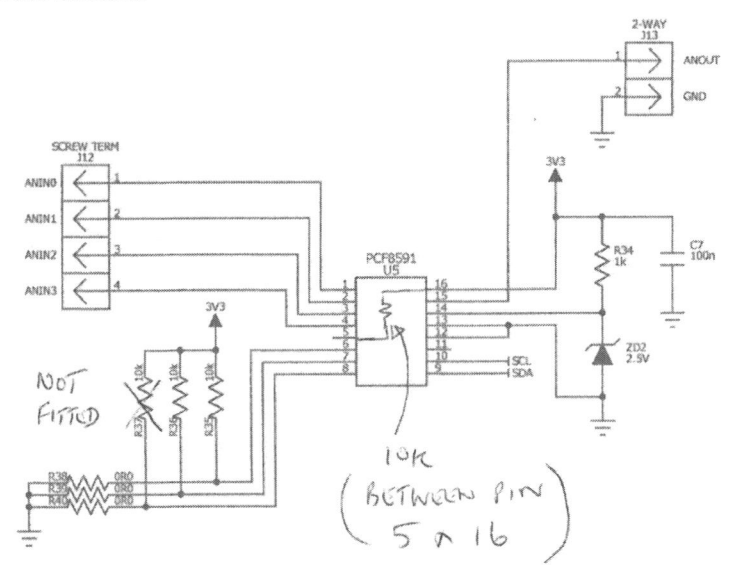

The PCF8591 provides 4 analogue inputs and 1 analogue output.

The I2C address is set by R35 and R36 pulling pins 6 and 7 high and is 0x4F. Pin 8 is not an I2C address pin and has to be pulled low through R40. R37 must not be fitted. Instead fit a 10k resistor between pins 5 and pins 16 of U5 to pull the 3rd I2C address pin high. (This is due to a mistake on the PCB layout where pins 5 and 8 were mixed up).

The information below shows how to set different I2C addresses for U5. Where a LINK is indicated a 0 ohm link is to be soldered into the PCB.

R38 fitted?	R39 fitted?	I2C address
NO	NO	0X49
NO	LINK	0x4B
LINK	NO	0x4D
LINK	LINK	0x4F

Note: The layout is for an axial zener, whereas we fit a LT1009 2.5V reference device which comes in a TO92 case. This is formed before fitting. With the flat side facing you, clip out the left hand side pin. Fit the central pin to the black line of ZD2 and the right hand side pin to the other (0V) side.

LCD drive circuit

A third MCP23008 (U4) device is fitted to drive an LCD. 4 of the pins are for the data and 3 are used for controlling the flow of data to the LCD. R31 and R33 are not fitted while R32 is a 0 ohm link - setting the I2C address as 0x25. Resistors R42 and R44 set the LCD contrast and the connections to pin 15 & 16 of the LCD is for the backlight.

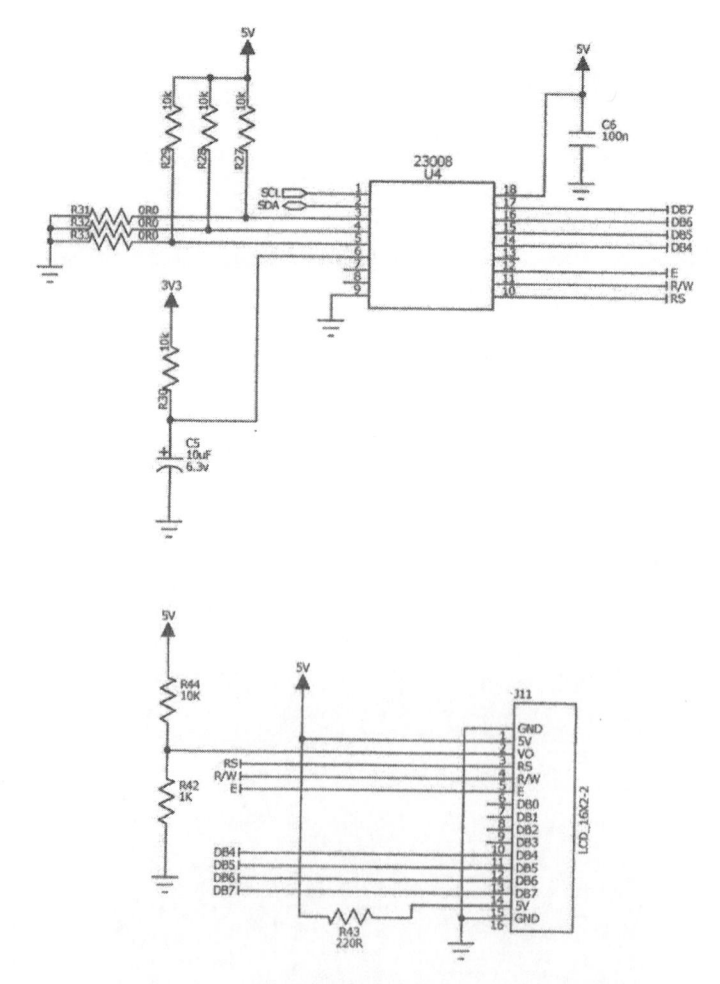

Using an MCP23008 to drive an LCD

The most common LCD controller IC in use is the Hitachi HD44780. This can be driven using 8 bits or 4 bits. In addition it uses 3 control pins which are Read/Write (R/W), Enable (E) and Register Select (RS). Using the 4-bit enables us to use a single 8-bit I2C IC (which provides 8 outputs) to drive the LCD, as we need a total of only 7 outputs.

The information below shows how to set different I2C addresses for U4. Where a LINK is indicated a 0 ohm link is to be soldered into the PCB.

R31 fitted?	R32 fitted?	R33 fitted?	I2C address
NO	NO	NO	0X27
NO	NO	LINK	0x26
NO	LINK	NO	0x25
NO	LINK	LINK	0x24
LINK	NO	NO	0x23
LINK	NO	LINK	0x22
LINK	LINK	NO	0x21
LINK	LINK	LINK	0x20

Raspberry Pi with Custard

7.3 Schematic

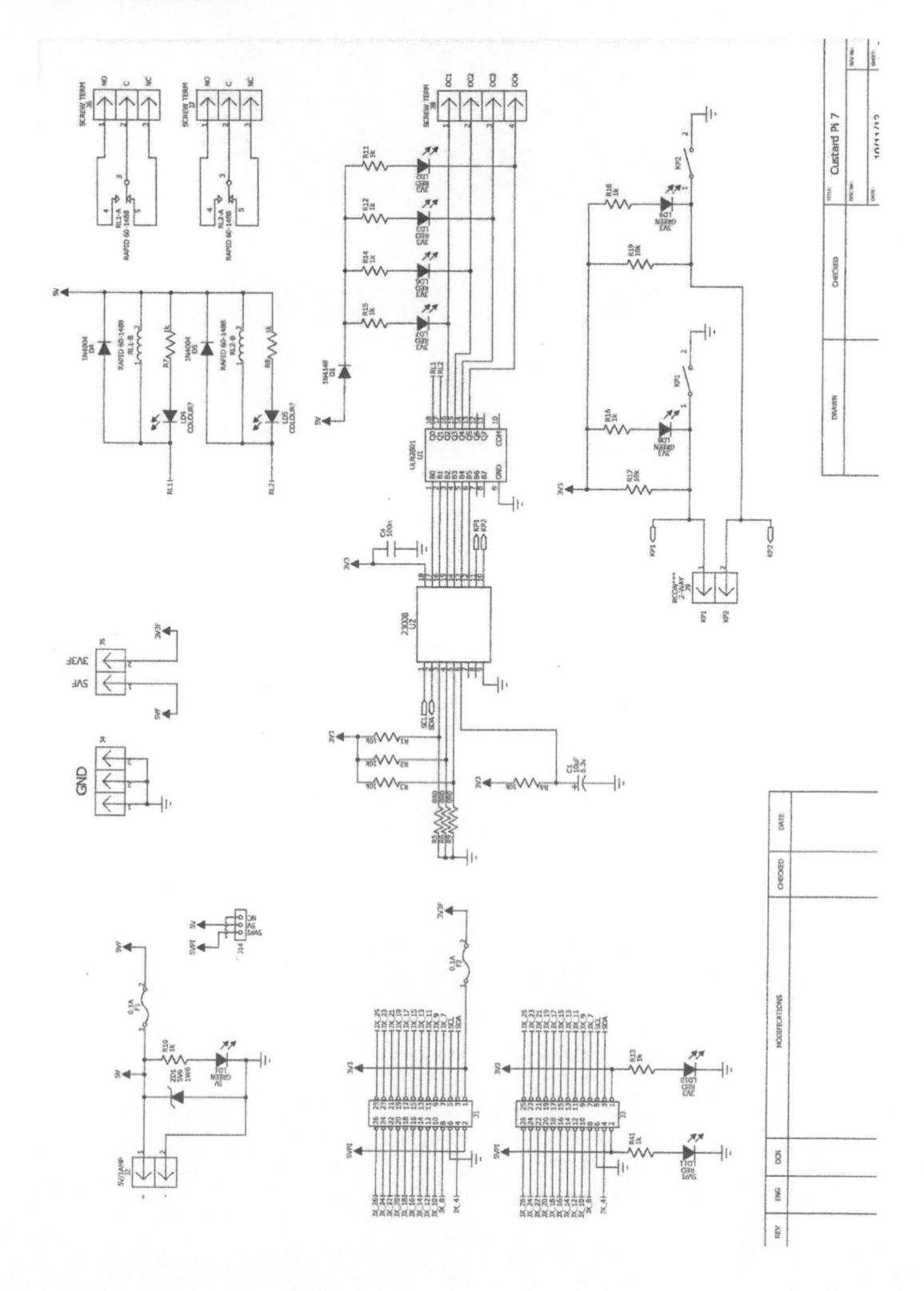

Raspberry Pi with Custard

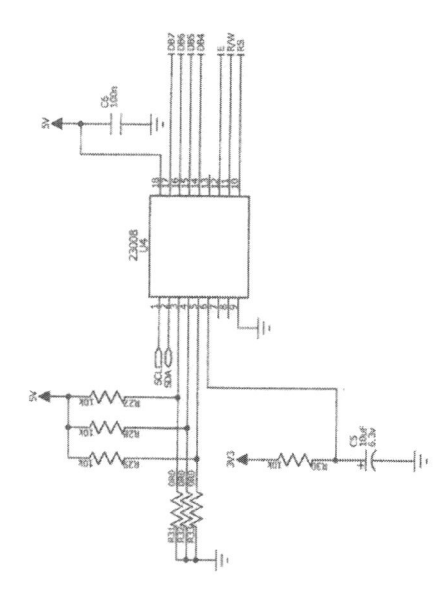

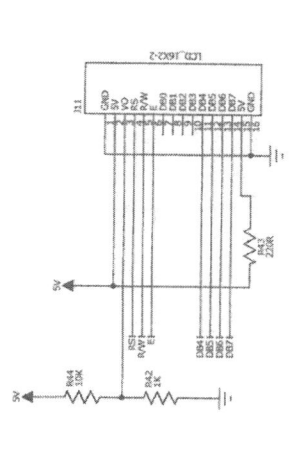

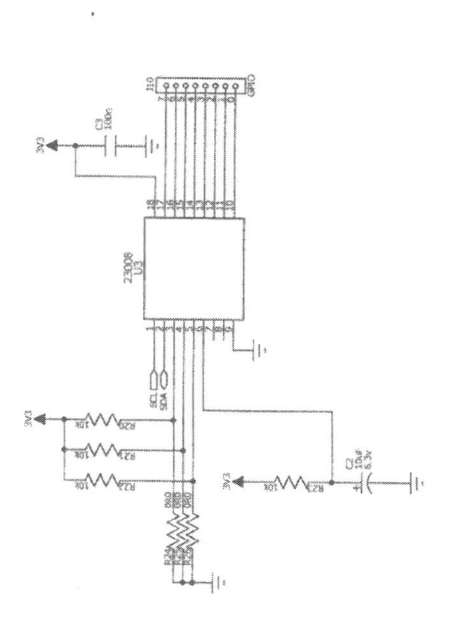

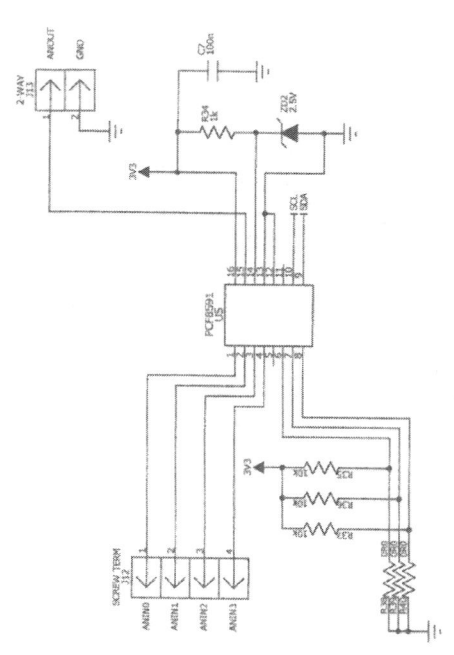

7.4 Testing the Relay and Open Collector outputs and Keypad inputs

First of all we need a driver routine that will help us to program the Custard Pi 7 quickly. The listing for this (cpi7x.py) is given below. This is similar to the Custard Pi 6 routines that we discussed in the previous chapter except for three differences.

1. Instead of 8 relays, we have 2 relays and 4 open collector outputs being switched ON and OFF.

2. As two of the pins are used as inputs, we have an additional 'readbit' command to read the status of the 2 keypad inputs.

3. Instead of setting all the pins as outputs, the 'setinandout' command sets 2 of the pins as inputs and the other 6 as outputs.

```
#1/usr/bin/env python
import time
import smbus

#***********************************************
# Custard Pi 7 resources v1.0 5th Dec 2013

#I2C addresses
#Set suitable address on Custard Pi 7

add0= 0x20
add1= 0x21
add2= 0x22
add3= 0x23
add4= 0x24
add5= 0x25
add6= 0x26
add7= 0x27

bus=smbus.SMBus(1)

#set IODIR register
iodir= 0x00
#set bit 0 and bit 1 as inputs, the others as outputs
inout= 0x03
#set GPIO register
gpio= 0x09
#set output latch
olat=0x0A

#set output ON
ONoc4= 0x04
ONoc3= 0x08
ONoc2= 0x10
```

Raspberry Pi with Custard

```python
ONoc1= 0x20
ONrl2= 0x40
ONrl1= 0x80

#set output OFF
OFFoc4= 0xFB
OFFoc3= 0xF7
OFFoc2= 0xEF
OFFoc1= 0xDF
OFFrl2= 0xBF
OFFrl1= 0x7F

def setbit(address, byte):
    #sets selected port pin
    outstatus = bus.read_byte_data(address, olat) | byte
    bus.write_byte_data(address, gpio, outstatus)

def clrbit(address, byte):
    #clears selected port pin
    outstatus = bus.read_byte_data(address, olat) & byte
    bus.write_byte_data (address, gpio, outstatus)

def readbit(address, bit):
    #read status of bit 0 or 1
    bitvalue = bus.read_byte_data(address, gpio) & bit
    return bitvalue

def setinandout(address):
    #set bit 0 and bit 1 as inputs, the others as outputs
    bus.write_byte_data(address, iodir, inout)

def alloff(address):
    #clear all relays
    bus.write_byte_data (address, gpio, 0x03)
```

#***

The test program below imports the routine above 'cpi7x' so that the commands described above can be used in the program. It sets the I2C address and initialises the card for 2 inputs and 6 outputs. Then it sist in a continuous loop switching the 4 Open collector outputs and the 2 Relay outputs ON and off with a 0.5 second in between.

Additionally, within each loop, the program checks the status of the keypad inputs and prints 'kp1 pressed' or 'kp2 pressed' if either of these are operated.

```python
#1/usr/bin/env python

import time
import cpi7x

#start program
```

Raspberry Pi with Custard

```python
board1=cpi7x.add7

cpi7x.setinandout(board1)

cpi7x.alloff(board1)

while True:
    cpi7x.setbit(board1, cpi7x.ONoc1)
    time.sleep (0.2)
    cpi7x.setbit(board1, cpi7x.ONoc2)
    time.sleep (0.2)
    cpi7x.setbit(board1, cpi7x.ONoc3)
    time.sleep (0.2)
    cpi7x.setbit(board1, cpi7x.ONoc4)
    time.sleep (0.2)
    cpi7x.setbit(board1, cpi7x.ONrl2)
    time.sleep (0.2)
    cpi7x.setbit(board1, cpi7x.ONrl1)

    time.sleep (0.5)

    cpi7x.clrbit(board1, cpi7x.OFFoc1)
    cpi7x.clrbit(board1, cpi7x.OFFoc2)
    cpi7x.clrbit(board1, cpi7x.OFFoc3)
    cpi7x.clrbit(board1, cpi7x.OFFoc4)
    cpi7x.clrbit(board1, cpi7x.OFFrl2)
    cpi7x.clrbit(board1, cpi7x.OFFrl1)

    time.sleep (0.5)

    kp2 = (cpi7x.readbit (board1, 0x01))
    kp1 = (cpi7x.readbit (board1, 0x02))

    if kp1 == 0:
        print "kp1 pressed"
    if kp2 == 0:
        print "kp2 pressed"

import sys
sys.exit()
```

7.5 Testing the General Purpose 8 Digital I/O

Any of these 8 bits can be set as inputs or outputs. In this section, we set 4 as outputs and 4 as inputs in order to show how to program them.

The driver routine for this (cpi7y.py) is similar to the other ones we have looked at already and is listed below. The exceptions are

1. The 'ONbit' and 'OFFbit' commands refer to the 4 bits that are used as outputs.

2. The 'setinandout' command sets the 4 least significant bits as inputs and the 4 most significant bits as outputs.

```
#1/usr/bin/env python
import time
import smbus

#***********************************************
# Custard Pi 7 resources v1.0 5th Dec 2013

#I2C addresses
#Set suitable address on Custard Pi 7

add0= 0x20
add1= 0x21
add2= 0x22
add3= 0x23
add4= 0x24
add5= 0x25
add6= 0x26
add7= 0x27

bus=smbus.SMBus(1)

#set IODIR register
iodir= 0x00
#set bit 0 to 3 as inputs,bits 4 to 7 as outputs
inout= 0x0F
#set GPIO register
gpio= 0x09
#set output latch
olat=0x0A

#set output HIGH
ONbit4= 0x10
ONbit5= 0x20
ONbit6= 0x40
ONbit7= 0x80

#set output LOW
OFFbit4= 0xEF
```

```
OFFbit5= 0xDF
OFFbit6= 0xBF
OFFbit7= 0x7F

def setbit(address, byte):
    #sets selected port pin
    outstatus = bus.read_byte_data(address, olat) | byte
    bus.write_byte_data(address, gpio, outstatus)

def clrbit(address, byte):
    #clears selected port pin
    outstatus = bus.read_byte_data(address, olat) & byte
    bus.write_byte_data (address, gpio, outstatus)

def readbit(address, bit):
    #read status of bit 0 to 3
    bitvalue = bus.read_byte_data(address, gpio) & bit
    return bitvalue

def setinandout(address):
    #set inputs & outputs
    bus.write_byte_data(address, iodir, inout)

def setpullups(address):
    #set inputs & outputs
    bus.write_byte_data(address, 0x06, 0xFF)

def alloff(address):
    #clear all outputs low
    bus.write_byte_data (address, gpio, 0x00)

#*******************************************
```

The test program imports 'cpi7y' in order to use the commands in that function. It then sits in an endless loop while switching the 4 outputs high and low and checking the status of the 4 inputs and printing the results on the screen.

```
#1/usr/bin/env python

import time
import cpi7y

#start program

board1=cpi7y.add6

cpi7y.setinandout(board1)

cpi7y.setpullups(board1)

cpi7y.alloff(board1)

while True:
```

```
    cpi7y.setbit(board1, cpi7y.ONbit4)
    time.sleep (0.2)
    cpi7y.setbit(board1, cpi7y.ONbit5)
    time.sleep (0.2)
    cpi7y.setbit(board1, cpi7y.ONbit6)
    time.sleep (0.2)
    cpi7y.setbit(board1, cpi7y.ONbit7)

    time.sleep (0.5)

    cpi7y.clrbit(board1, cpi7y.OFFbit4)
    cpi7y.clrbit(board1, cpi7y.OFFbit5)
    cpi7y.clrbit(board1, cpi7y.OFFbit6)
    cpi7y.clrbit(board1, cpi7y.OFFbit7)

    time.sleep (0.5)

    bit0 = (cpi7y.readbit (board1, 0x01))
    bit1 = (cpi7y.readbit (board1, 0x02))
    bit2 = (cpi7y.readbit (board1, 0x04))
    bit3 = (cpi7y.readbit (board1, 0x08))

    if bit0 == 0:
        print "bit0 low"
    else:
        print "bit0 high"

    if bit1 == 0:
        print "bit1 low"
    else:
        print "bit1 high"

    if bit2 == 0:
        print "bit2 low"
    else:
        print "bit2 high"

    if bit3 == 0:
        print "bit3 low"
    else:
        print "bit3 high"

    print "***************************"
import sys
sys.exit()
```

7.6 Testing the LCD interface

First of all we need to prepare an LCD with a suitable connector. The connections provided on the Custard Pi 7 for the LCD is shown below.

Raspberry Pi with Custard

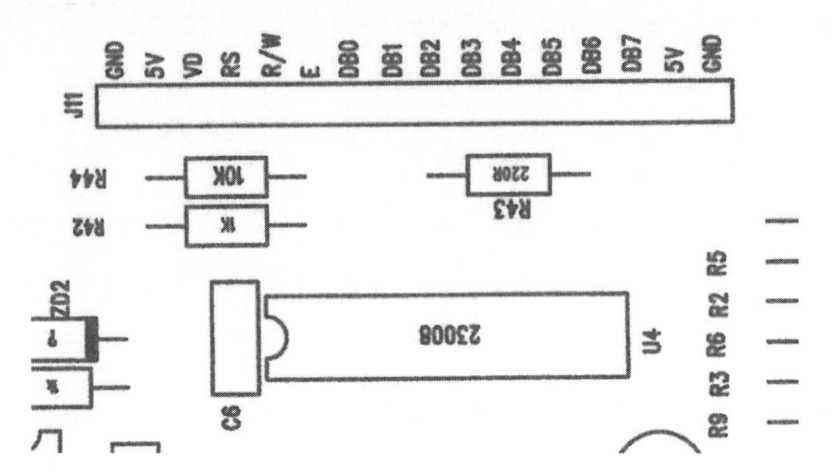

The connections on a 2 line LCD display (using a Hitachi HD44780 controller) are shown below. From this one can see that a 16 way connector can be used to connect the two together very easily.

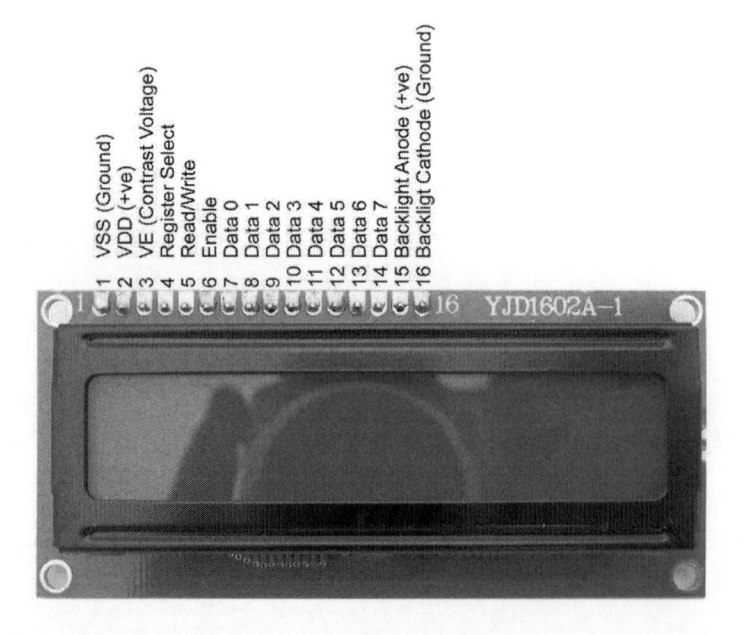

Once the LCD has been plugged, we are ready to have a look at the driver routine required (cpi7lcd.py) to write to the LCD.

```
#1/usr/bin/env python
import smbus
from time import sleep

port = 1
bus = smbus.SMBus(port)

def write(byte):
  bus.write_byte_data(addr, 0x09, byte)

def read():
```

```python
    return bus.read_byte_data(addr, 0x0A)

def read_nbytes_data(data, n): # For sequential reads > 1 byte
    return bus.read_i2c_block_data(addr, data, n)

def lcdinit(address):
    global addr
    addr=address
    #set IODIR register
    iodir= 0x00
    #set GPIO register
    gpio= 0x09
    #set all 8 bits as outputs
    bus.write_byte_data(addr, iodir, 0x00)

    delay=0.0005

    write(0x30)
    lcd_strobe()
    sleep(delay)
    lcd_strobe()
    sleep(delay)
    lcd_strobe()
    sleep(delay)
    write(0x20)
    lcd_strobe()
    sleep(delay)

    lcd_write(0x28)
    lcd_write(0x08)
    lcd_write(0x01)
    lcd_write(0x06)
    lcd_write(0x0C)
    lcd_write(0x0F)

# clocks EN to latch command

def lcd_strobe():
    write((read() | 0x04))
    write((read() & 0xFB))

# write a command to lcd
def lcd_write(cmd):
    write((cmd >>4) <<4)
    lcd_strobe()
    write((cmd & 0x0F)<<4)
    lcd_strobe()
    write(0x0)

# write a character to lcd (or character rom)
def lcd_write_char(charvalue):
    write((0x01 | (charvalue >>4)<<4))
    lcd_strobe()
    write((0x01 | (charvalue & 0x0F)<<4))
    lcd_strobe()
    write(0x0)
```

```python
# put char function
def lcd_putc(char):
    lcd_write_char(ord(char))

# put string function
def lcd_puts(string, line):
  if line == 1:
   lcd_write(0x80)
  if line == 2:
   lcd_write(0xC0)
  if line == 3:
   lcd_write(0x94)
  if line == 4:
   lcd_write(0xD4)

  for char in string:
   lcd_putc(char)

# clear lcd and set to home
def lcd_clear():
  lcd_write(0x01)
  lcd_write(0x02)
```

Once the driver routine has been loaded onto the SD card one can run the test program to write characters to the LCD. This imports the function 'cpi7lcd' and then sets the I2C address to '0x25' before writing to both lines of the display.

```python
#1/usr/bin/env python

import time
import cpi7lcd

#start program

lcd = cpi7lcd

lcd.lcdinit(0x25)

while True:
    lcd.lcd_clear
    time.sleep (0.5)
    lcd.lcd_puts("LCD line 1 test",1)   #display text on line 1
    time.sleep (0.5)
    lcd.lcd_clear
    lcd.lcd_puts("LCD line 2 test",2)   #display text on line 1
    time.sleep (0.5)
    lcd.lcdinit(0x25)

import sys
sys.exit()
```

7.7 Testing the Analogue inputs and outputs

It is fairly easy to communicate with the PCF8591 IC which handles the 4 analogue inputs and single analogue output so there is no seperate function for this. The program below has 2 functions.

'readanalog' reads in turn from each of the analogue input channels. The function 'writeanalog' sends an 8 digit value to the analogue output channel.

The program sets the I2C address to '0x4F' and defines the 4 channels. However it uses the auto-increment feature to read from each channel in turn. It presents a value from 0 up to 255 to the analogue output channel. When the program is running, simply connect a test lead from the analogue output to each of the analogue inputs in turn to check that the correct value is being measured and printed to the screen.

Note: The value received back might be a digit out due to measurement tolerances.

```
#1/usr/bin/env python
import time
import smbus

#************************************************
# Custard Pi 7 8591 25th Oct 2013

#I2C addresses

add= 0x4F
bus=smbus.SMBus(1)

#DtoA channel select
ch0= 0x00
ch1= 0x01
ch2= 0x02
ch3= 0x03

def readanalog(add):
#    bus.write_byte_data(add,ch,0x00)
    analog=bus.read_byte(add)
    return analog

def writedtoa(add,value):
    bus.write_byte_data (add, 0x44, value)

#************************************************
while True:
    value=0x00
    for x in range (0,255):
        writedtoa(add,value)
```

```
    print value
    time.sleep(0.5)
    value=value+1

    an0 = readanalog(add)

    an0 = readanalog(add)
    print "ch0=", (an0)

    an1 = readanalog(add)
    print "ch1=", (an1)

    an2 = readanalog(add)
    print "ch2=", (an2)

    an3 = readanalog(add)
    print "ch3=", (an3)

    print "*****"

    time.sleep(0.5)
```

CHAPTER 8   CUSTARD PI 8
Surface Mount and Through Hole Prototyping Board

Many chips are only available as a Surface Mounted Device (SMD). However it is not easy to build prototypes using SMD components. Most of the prototyping options available are for through-hole components. There are adaptor boards that will allow an engineer to try out a SMT integrated circuit but they are limited in scope.

The Custard Pi 8 is a flexible prototyping board that can be used for SMD as well as through-hole electronic components.

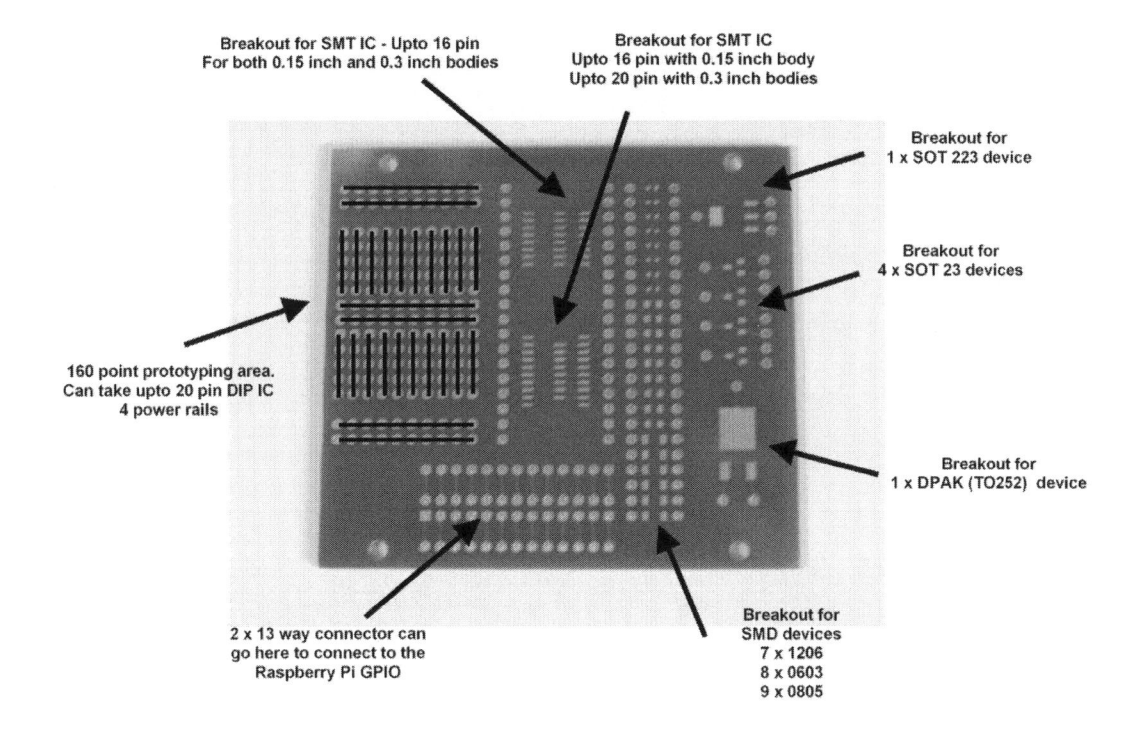

It can be quickly connected to the Raspberry Pi using a ribbon cable and a 2x13 way connector. In this way, it can be used as a breakout board.

The images below show the circuit for the electronic dice project from Chapter 2 built up using the CPi8 and surface mount resistors. As we are not using the Custard Pi 8, we have to use resisotrs to limit the current flowing through the LEDs of the 7-segment display.

CHAPTER 9   PI CASING AND PI BASE

The Pi Casing will enclose the Raspberry Pi with a variety of GPIO accessories plugged in. It will also allow prototyping boards to sit alongside the Pi for custom hardware.

The space on the right hand side and the mounting holes are designed to take the Custard Pi 8 as shown below.

Alternatively a breadboard can be fitted as shown here.

Note: The Pi Casing has mounting holes that is compatible with the Model B and the Model B+ versions of the Raspberry Pi.

The Pi Base provides a stable mounting plate for the Raspberry Pi and any a prototyping board. It does not have a covering plate like the Pi Casing.

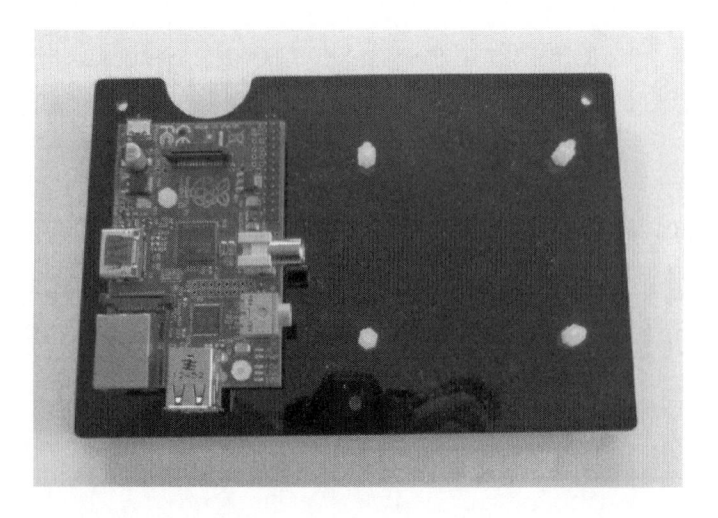

CHAPTER 10   CUSTARD PI COMBOS
Mixing and Matching Custard Pi's

Introduction

If a particular Custard Pi does not have all the right inputs and outputs, then it's possible to combine more than one. This chapter looks at the various combinations that are possible.

GPIO bus connection method

The various cards use different means of connecting to the Raspberry Pi GPIO and this summarised below.

SPI bus	I2C bus	8 digital I/O
Custard Pi 2	Custard Pi 6	Custard Pi 1
Custard Pi 3	Custard Pi 7	Custard Pi 2
		Custard Pi 5

The card combinations are limited by the type of bus connection used. For example, the SPI bus can only address 2 devices. As the Custard Pi 2 has a Analogue to Digital Convertor (ADC) and a Digital to Analogue Convertor (DAC) and both of these use the SPI bus, you cannot have more than 1 Custard Pi 2 connected to the GPIO. Because the Custard Pi 3 also uses the SPI bus, you cannot have a Custard Pi 2 and a Custard Pi 3 connected to the Raspberry Pi at the same time.

The I2C bus is addressable and you can have as many I2C devices as there are addresses. This means that you can have up to 8 Custard Pi 6's connected at the same time. You just have to make sure that the switches on each card are used to set a different I2C address.

Example 1 - 64 relays controlled from the Raspberry Pi

Eight Custard Pi 6 cards daisy chained to provide a total of 64 relays that can be controlled from the Raspberry Pi GPIO.

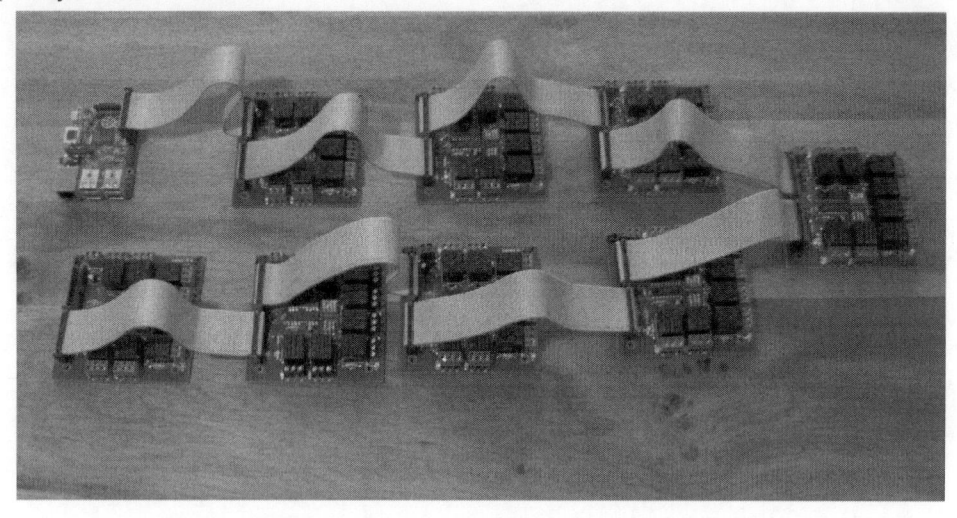

Example 2 - 8 relays and 8 analogue inputs

This is achieved by using a Custard Pi 6 and a Custard Pi 3. As one uses the I2C bus and the other uses the SPI bus, this is possible.

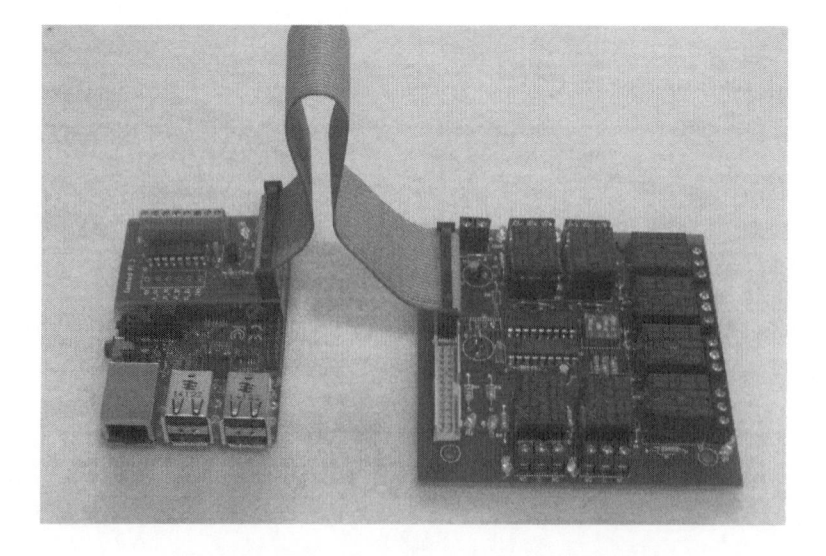

Example 3 - 8 relays, 8 analogue inputs and 8 digital I/O

This is achieved by using a Custard Pi 6, Custard Pi 3 and a Custard Pi 5 as shown below.

Raspberry Pi with Custard

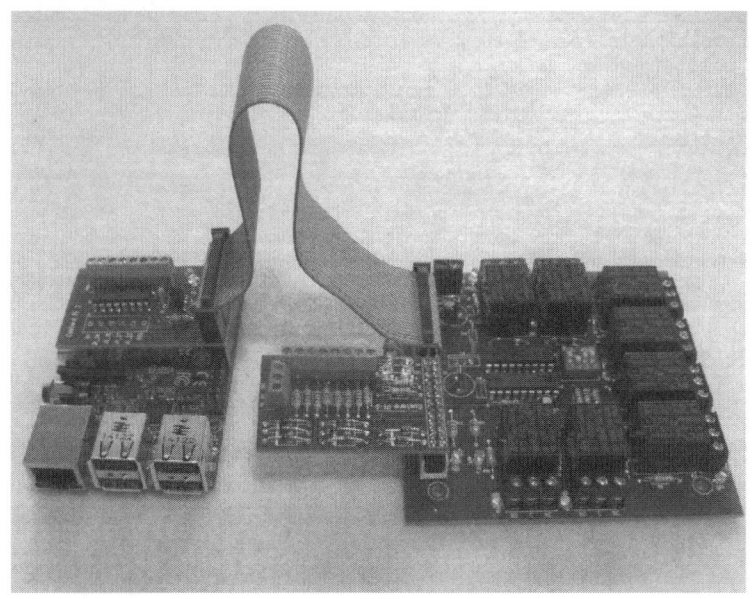

Example 4 - 8 relays, 2 analogue I/P, 2 analogue O/P, 4 open collector outputs and 4 digital inputs

This is easily achieved by combining the Custard Pi 6 and the Custard Pi 2.

Example 5 - 8 Analogue inputs and 8 digital I/O

Plugging in the Custard Pi 3 and Custard Pi 5 into the Raspberry Pi GPIO will provide these functions.

Example 6 - Custard Pi 7 with 12 bit analogue inputs

The Custard Pi 7 has 4 analogue inputs but these are only of 8 bit resolution. If a particular application requires a higher resolution, then the quickest way to achieve this is to plug in a Custard Pi 3 card into the Custard Pi 7.

Other Custard Pi 7 combinations

The photos below show the Custard Pi 7 with the Custard Pi 2, Custard Pi 5 and also with both the Custard Pi 3 and 5.

Custard Pi 7 with Custard Pi 2

Custard Pi 7 with CPi5

Custard Pi 7 with CPi5 and CPi3

Physical connection method

The Custard Pi 3 and 5 have stacking connectors. This means that they can be both plugged into the Raspberry Pi GPIO at the same time. The Custard Pi 6 and 7 have two ribbon cable sockets - again these can be used for 'daisy-chaining'. However the Custard Pi 1 and 2 cannot be extended upon if they are the first card to be plugged in. If you want to combine these cards with others they have to be the last in the chain.

APPENDIX A   STARTING UP DESKTOP ON POWERUP

From the command line prompt or LXTerminal, type "`sudo raspi-config`". You will get the screen below. For boot_behaviour select desktop. When you next power up, the Raspberry Pi will load up the GUI automatically (without you having to type in "startx").

```
Raspi-config

    info              Information about this tool
    expand_rootfs     Expand root partition to fill SD card
    overscan          Change overscan
    configure_keyboard Set keyboard layout
    change_pass       Change password for 'pi' user
    change_locale     Set locale
    change_timezone   Set timezone
    memory_split      Change memory split
    ssh               Enable or disable ssh server
    boot_behaviour    Start desktop on boot?
    update            Try to upgrade raspi-config



            <Select>                    <Finish>
```

APPENDIX B   AUTOSTARTING A PYTHON PROGRAM ON POWERUP

First code and debug your Python program. Make sure that the Pi starts up desktop on power up, as described in Appendix A.

Then we need to edit /etc/rc.local by typing

```
sudo nano /etc/rc.local
```

Note: Nano is a command line editor. You navigate round the text by using the cursor keys or CTRL key commands which are summarised at the bottom of the screen.

At the bottom, just above exit 0 we'll add a call to our script.

```
sudo python /home/pi/yourprogram.py
```

Save your changes. Now every time you power up, "yourprogram.py" will run.

APPENDIX C   COMPARISION   OF   RASPBERRY   PI MODELS

Since it's introduction there have been various versions of Raspberry Pi made available. Between revision 1 and revision 2 boards some of the GPIO board pins changed. This is shown in the chart below.

Rev 2	Rev 1	Function	Pin Numbers		Function	Rev 1	Rev 2
		Power 3.3 V	1	2	Power 5V		
		I2C SDA	3	4	***	DNC	5V
		I2C SCL	5	6	Gnd		
		GPIO 4	7	8	UART TXD		
Gnd	DNC	***	9	10	UART RXD		
		GPIO 17	11	12	GPIO 18		
GPIO 27	GPIO 21	***	13	14	***	DNC	Gnd
		GPIO 22	15	16	GPIO 23		
3.3V	DNC	***	17	18	GPIO 24		
		SPI MOSI	19	20	***	DNC	Gnd
		SPI MISO	21	22	GPIO 25		
		SPI CLK	23	24	SPI CE0		
Gnd	DNC	***	25	26	SPI CE1		

How can you tell if you have a revision 1 or a revision 2 board? The easiest way is to execute the following commands.

```
sudo python
import RPi.GPIO as GPIO
GPIO.RPI_REVISION
```

The value returned will indicate the type of board in use.

0 = Compute Module
1 = Revision 1 board
2 = Revsions 2 board
3 = Model B+/A+

The main differences between Model A, B, A+ and B+ are summarised in the table below.

	Model A	Model A+	Model B	Model B+
Memory	256 MB	256 MB	512 MB	512 MB
USB Ports	1	1	2	4
Storage	SD card	Micro SD	SD card	Micro SD
Ethernet	None	None	Yes	Yes
Digital I/O	8	17	8	17
SPI	yes	yes	yes	yes
I2C	yes	yes	yes	yes
UART	yes	yes	yes	yes
EEPROM ID	no	yes	no	yes

Printed in Great Britain
by Amazon